Principles of
Green Bioethics

Principles of
Green Bioethics

SUSTAINABILITY IN HEALTH CARE

Cristina Richie

Michigan State University Press | East Lansing

♾ The paper used in this publication meets the minimum requirements of
ANSI/NISO Z39.48-1992 (R 1997) (Permanence of Paper).

Michigan State University Press
East Lansing, Michigan 48823-5245

Printed and bound in the United States of America.

28 27 26 25 24 23 22 21 20 19   1 2 3 4 5 6 7 8 9 10

LIBRARY OF CONGRESS CATALOGING-IN-PUBLICATION DATA
Names: Richie, Cristina, author.
Title: Principles of green bioethics : sustainability in health care / Cristina Richie.
Description: East Lansing : Michigan State University Press, [2019] |
Includes bibliographical references and index.
Identifiers: LCCN 2018047545| ISBN 9781611863239 (pbk. : alk. paper) | ISBN
9781609176020 (pdf) | ISBN 9781628953688 (epub) | ISBN 9781628963694 (kindle)
Subjects: LCSH: Bioethics. | Medical care. | Green movement. | Environmentalism.
Classification: LCC QH332 .R533 2019 | DDC 174.2—dc23
LC record available at https://lccn.loc.gov/2018047545

Book design by Charlie Sharp, Sharp Designs, East Lansing, MI
Cover design by Shaun Allshouse, www.shaunallshouse.com

Michigan State University Press is a member of the Green Press Initiative and is
committed to developing and encouraging ecologically responsible publishing
practices. For more information about the Green Press Initiative and the use of
recycled paper in book publishing, please visit *www.greenpressinitiative.org*.

Visit Michigan State University Press at *www.msupress.org*

....

This book is dedicated
to the Ethics faculty
in the Boston College Department of Theology
and Boston College School of Theology and Ministry,
particularly James Keenan, Andrea Vicini,
and Lisa Cahill

....

Contents

Foreword

Paul R. Ehrlich

have spent decades thinking about bioethics and quickly accepted when Dr. Cristina Richie asked me to write this foreword for her fine manuscript, *Principles of Green Bioethics: Sustainability in Health Care*. As I expected, it's been a real pleasure, as Dr. Richie covers in depth, and with interesting examples, a vast array of bioethical issues related to health care and sustainability, including assisted reproduction, sex reassignment surgery, medical treatment of obesity, and oncofertility. Her book takes a look at the future of telemedicine and teleclinics. She distinguishes and discusses different kinds of equality. She points out some of the costs of racism in the medical system and covers tough subjects like the profit motive among other unfortunate aspects of the medical industrial complex. *Green Bioethics* includes advice for green patients, green doctors, and green health insurance. And it often brings up subjects and viewpoints that I had not run into before, despite having written on bioethics.

It was pure coincidence that when Dr. Richie contacted me, I had just completed my first detailed study in an area of health care. Dr. Sandra Kahn, a leading clinical orthodontist, and I examined the environmental impacts of the "jaws epidemic": the effects of industrialization on the evolution of the human jaw and the impact of that jaw evolution on human well-being. With Dr. Richie's book for guidance, writing this foreword gave me the opportunity to apply the principles of Green Bioethics to cosmetic dentistry. The environmental-ethical issues raised by Dr. Richie on how civilization should deal with the health of our entire bodies is also a pertinent guide to how civilization should deal with the changes in our jaws—an area of increasing health-care concern that has been artificially divided from the health of our entire bodies.

Richie's first principle of Green Bioethics—distributive justice: allocate basic medical resources before special-interest access—is of course drowned in the general lack, indeed in the general decline, of distributive justice in all areas of health care today. Ending the jaws epidemic could go a long way toward satisfying that first principle. Braces, wisdom tooth extraction, continuous positive airway pressure (CPAP) machines, jaw surgery, and the like are environmentally costly. They are also financially costly and thus less available to poor portions of populations. It is also those population segments that are most likely to suffer the impacts of environmental deterioration.

Richie's second principle of Green Bioethics—resource conservation: provide health-care needs before health-care wants—is obviously connected with plastic surgery, and it is also present in oral treatments ranging from braces, extractions, cosmetic tooth bleaching, dental implants, and surgery done to improve appearance. As Richie says, "doctors who wish to employ resource conservation could halt medical developments, techniques, and procedures that are non-clinically beneficial." Improved appearance is not "clinically beneficial"—especially when it may lead to a more restricted airway and sleep apnea.

Richie's third principle of Green Bioethics—simplicity: reduce dependence on medical interventions—is highlighted in the jaws epidemic, where a stunning saving of medical resources can clearly be made by changing from treatment to prevention. Parents can do much to change the MacDonald's environment—a diet of soft, processed foods that give jaw muscles little exercise—that is a fundamental cause of the problem. They can also help their children to exercise their jaws and maintain the oral posture that will lead to proper jaw development. So can actions by society, such as considering the need for "chewability" when making dietary rules and recommendations, reducing the need for medical treatment, which is the essence of simplicity. Moreover, preventing the jaws epidemic would save medical resources and some of the energy and materials used in processing foods, just as a reduction in smoking avoids the erosion of soil from growing tobacco, and the energy and materials used in growing, manufacturing, shipping, and marketing.

Dr. Richie's fourth principle of Green Bioethics—ethical economics: humanistic health care instead of financial profit—concerns the frequent conflicts between health-care service and financial incentives. About 20 million "wisdom teeth" are chopped out of 5 million mouths annually in the United States, at a cost of about $3 billion. Those operations result in more than 11 million patient days of discomfort postoperatively, and more than 11,000 people suffer permanent paresthesia—numbness of the lip, tongue, and cheek—as a consequence of nerve injury during the surgery. At least two thirds of these extractions, associated costs, and injuries are unnecessary, constituting a silent epidemic of medically caused injury that afflicts tens of thousands of people with lifelong discomfort and disability.

The principles of Green Bioethics have stimulated me to continue to think about environmental ethics and health care. We need much more public discussion of the ethical dimensions of the human predicament of massive resource use, climate change, and unequal health care. This thought-provoking book should help stimulate it.

Acknowledgments

The publication of this book would not have come about without the support, encouragement, and tutelage of a number of people and organizations.

First, and foremost, I am deeply appreciative to my colleagues in the Bioethics and Interdisciplinary Studies Department, in the Brody School of Medicine at East Carolina University. To Maria Clay, the chair, for her leadership and organization that permitted me the time to write. To Greg Hassler, who read every word of my manuscript with alacrity and offered insightful comments. To Kenneth DeVille and Todd Savitt, who were valuable guides through the publication process. And, to Sheena Eagan, who kept me accountable with my revisions.

To the Ethics faculty in the Boston College Department of Theology and Boston College School of Theology and Ministry, to whom this book is dedicated. To James Keenan, who has driven my work to new heights; Andrea Vicini, who has been a mentor and sage; and Lisa Cahill, whose vision for global health care has been a major influence on my intellectual development. To Mary Ann

Hinsdale, who always had an edifying word for me, and James Bretzke, Kenneth Himes, and Mary Jo Iozzio, who live social justice as they teach it.

I am grateful to my editor, Julie Loehr, who has made the process enjoyable.

Conceptually, Green Bioethics and the principles therein have appeared in previously published material. I acknowledge, with deep appreciation, the journals that have permitted me republication. In 2012, I first put pen to paper and submitted a short essay on Green Bioethics to the Catholic Health Association. It was selected for the Catholic Health Association Annual Theology and Ethics Colloquium Award in 2013. I was honored to receive this recognition, and credit the early support of the CHA—particularly Ron Hammel and Tom Nair—as a major factor in developing Green Bioethics into a full monograph. The original article, "Building a Framework for Green Bioethics: Integrating Ecology into Health Care," was first published in the Fall 2013 edition of *Health Care Ethics USA* by the Catholic Health Association of the United States. Portions of chapter 3 appeared in "Global Justice and Maternal Resources: Taking a Note from Catholic Social Teachings," *Developing World Bioethics* 15, no. 3 (2015): 179–90.

Finally, I have enjoyed the friendship of a number of good souls who have made the creation of this book a delight. Stas, thank you for your support while I worked on this book over two years, in three U.S. cities and five countries worldwide. Your editorial wisdom, good humor, and personal attentiveness is much appreciated. Jen, your wit, kindness, and intelligence has abided with me since graduate school and I admire your drive. Sheena, your presence has been the perfect balance of inspiration and competition, grit and tranquility.

Gratitude is a panacea for many of the ills of the modern world. May we all cultivate more of it.

Introduction

Concern over the deterioration of the environment has been a significant feature of ethics over the last half century. This complex and multifaceted quagmire continues to demand global attention even as policies to curtail climate change lag. Talk of "climate chaos"—the natural disasters that result from climate change—looms large in news coverage. The effects of human beings on the environment are so extensive that we are now said to be living in the era of the Anthropocene, which is defined by "human influence on Earth, where we have become a geological force shaping the global landscape and evolution of our planet."[1] This ignoble definition is characterized by two interconnected challenges: human population growth and resource consumption.

Human Population Growth

In the last one hundred years, there has been a drastic increase in the number of human inhabitants in the world, caused by an explosion in reproduction rates in tandem with prolonged lifespans. This trend is expected to continue. The United Nations Population Division's recent 2012 *World Population Prospect* predicted that with medium fertility rates—2.53 children per woman—the world's population would reach 8.1 billion in 2025, increase to 9.6 billion in 2050, and peak at 10.9 billion by 2100.[2] Population growth projections do not account for externalities such as competition for water, food, land, living space, and the fallout of such competition that might result in poverty, war, and stress on the ecosystem.

Furthermore, these numbers cannot fully account for the impact that human population growth has on the nonhuman community. Biologists estimate that at least one thousand plant and animal species become extinct annually. Many of these species are lost as a direct result of human development. And, due to current population-growth trajectories, "even if the human collective were to pull as hard as possible on the total fertility policy lever . . . the result would be ineffective in mitigating the immediately looming global sustainability crises (including anthropogenic climate disruption), for which we need to have major solutions well under way by 2050 and essentially solved by 2100,"[3] according to a report from the *Proceedings of the National Academy of Sciences of the United States of America*. The ability of the earth to sustainably provide for all of its inhabitants is impacted by human resource consumption as well as human population growth.

Human Resource Consumption

Food, water, energy, lumber, minerals, and vegetation are fundamental to maintain elementary human life, but the drive to consume more than what is essential is ravaging the planet. Even basic resource use can be exploitative if the manner in which they are derived is overly consumptive. For instance, fresh water is necessary for all life. In developed countries, tap water is clean, affordable, and readily available. However, many people, businesses, and organizations—including health-care facilities—choose to purchase and consume bottled water. In 2007, over 200 billion liters of bottled water were sold, mostly in the United States and other countries where bottled water is unnecessary to ensure cleanliness.[4] This extravagance is an environmental burden on multiple other resources.

Individual plastic water bottles are ubiquitous in America and other developed countries. In order to obtain bottled water, fresh water must first be gathered from a location, purified, and then transported to a bottling factory. Oil must be extracted to fuel bottle manufacturing, which depends on petroleum and plastic in order to create the disposable bottles. After bottles have been made and the water has been transported to the factory, energy is used—typically from nonrenewable sources—for the machinery to bottle the water. Once the water is in the bottle, a label using paper and ink is placed on each individual container; bottles are then grouped and packaged for transportation, using cardboard and more plastic for wrapping, and then shipped via truck or plane, which uses fossil fuel as well.

These individually wrapped, individually labeled bottles are delivered to hospitals, among other businesses, and then stored in coolers that require energy, even though water does not need to be cold to avoid spoiling. The Pacific Institute estimated that "approximately 17 million barrels of oil equivalent were needed to produce the plastic water bottles consumed by Americans in 2006— enough energy to fuel more than one million cars for a year."[5] This illustrates the profligate waste of just one consumer practice. Water bottles are sold alongside a number of other prepackaged food and beverage products that follow similar consumptive lines, while simultaneously being sold or distributed by people or organizations that are relying on buildings, materials, energy, and massive resource use for their infrastructure.

Resource consumption has a second component. In addition to upstream resource use, which identifies what is being used (e.g., water) and how (e.g., in disposable bottles), downstream resource use calculates the carbon emissions after a product has been consumed. Environmentalists therefore speak of the ecological crisis in terms of initial, upstream resource consumption and final, downstream use, that is, carbon dioxide (CO_2) emissions. Products as well as individuals, families, countries, and the world aggregate have a carbon impact.

For instance, in 2015, China and the United States had the highest CO_2 emissions output, with China emitting 10,641,788.99 ktons and the United States 5,172,337.73 ktons.[6] To put this in perspective, the entire European Union, with twenty-eight member states, ranked third in carbon emissions at 3,469,670.82 ktons.[7] Unfortunately, the carbon emissions of individual countries do not stay within national borders. While industrialized countries prosper from resource consumption while dually polluting the environment, people in developing countries suffer disproportionately from poverty and rapid climate change.

Human population growth and human resource consumption are twin environmental issues facing the world. Data overwhelmingly supports the urgency of reducing population growth and consumption. At this critical moment in human history, where our future depends on our ability to adapt and address the environmental perils of our modern society, all areas of human life must be analyzed for their ability to conform to ecological priorities. Health care is no exception.

Health Care and Climate Change

Health care is ubiquitous in the lives of industrialized people. Yet, every medical development, technique, and procedure impacts the environment. In the last decade, the health-care sector in both the United States and United Kingdom were assessed to be significant contributors to resource consumption and carbon dioxide emissions. Since carbon dioxide creates climate change–related health hazards, health care not only contributes to climate change but must also care for those affected by climate change. This Sisyphean cycle has been noted by bioethicists, among others.

The Impact of Health Care on Climate Change

Carbon output is partially determined, among other factors, by country of locale. Within health care, this point is especially germane. When buildings, energy, food, medical devices, drugs, and follow-up care are all accounted for, the very nature of medical services rendered in American hospitals are more carbon-intensive than those in other countries. The *Journal of the American Medical Association* (*JAMA*) featured an article assessing the carbon output of health care in the United States. The United States medical industry is responsible for nearly a tenth of the country's carbon dioxide emissions.[8] This statistic would be inconsequential if the U.S. had a sustainable carbon output. However, the United States is among the top polluting countries in the world. Thus, the carbon cost from using health care in America is among the highest in the world. In 2007, the U.S. health-care sector expended an estimated 546 million metric tons of carbon dioxide.[9] *JAMA* furthermore reported that the largest contributors to carbon emissions in the health-care sector were the hospital and prescription-drug sectors, at 39 percent and 14 percent, respectively.

It should be self-evident that hospitals in the United States have large carbon footprints since they are habitually outfitted with air conditioning for comfort, rely on subzero temperatures for storage of certain products, and require consistent energy, as well as backup generators for machines that sustain life functioning. It is unexpected, perhaps, that pharmaceuticals would be a substantial environmental menace. Yet, the resources involved in research, development, manufacturing, and distribution of drugs result in a sizable carbon output. This data is reflected in other health-care systems as well.

The United Kingdom examined the carbon expenditure of its National Health Service (NHS) and calculated that the NHS is responsible for 18 million tons of carbon dioxide each year. This figure is thirty times *less* than U.S. health-care carbon emissions. Interestingly, the carbon from National Health Service emissions is not from familiar high-carbon culprits like meat in cafeterias, facilities that rely on electricity and air conditioning, or even single-use instrument waste. Rather, "pharmaceuticals contribute most to procurement emissions, being responsible for four million tons of carbon per year."[10] The NHS accounts for 25 percent of England's total public-sector emissions, and is a major driving force in climate change.

The Impact of Climate Change on Health Care

In addition to contributing to carbon emissions, health care must also care for those who suffer from climate change–related health problems. It is well documented that climate change causes health problems. The World Health Organization records that worldwide climate change presents "potential risks to health [that] include deaths from thermal extremes and weather disasters, vector-borne diseases, a higher incidence of food-related and waterborne infections, photochemical air pollutants and conflict over depleted natural resources."[11] Climate change–related health hazards are an international issue with unique domestic contours.

Within the United States, six specific climate change–related events—national ozone air pollution from 2000 to 2002, the West Nile virus outbreak in Louisiana in 2002, the Southern California wildfires in 2003, the Florida hurricane season in 2004, the California heat wave in 2006, and the Red River flooding in North Dakota in 2009—caused more than 760,000 encounters with the health-care system and over $740 million in health costs.[12] Complications from climate change, including food scarcity, respiratory disease, and drought, will continue to add to the burdens of health care for years to come.[13]

The future of our world—and our own personal health—may very well depend on how effectively we halt ecological destruction and conserve our resources. As such, this book will propose Green Bioethics, a principled framework for evaluating the sustainability of medical developments, techniques, and procedures. The four principles of Green Bioethics integrate the scope of Western biomedical ethics with the concerns of environmental ethics, resulting in a conservation-based bioethic that will reduce resource consumption in health care and attenuate climate change–related health hazards. The planet cannot afford to have ecological ethics and biomedical ethics remain distinct fields. Green Bioethics offers a practical philosophy capable of significantly transforming ethics.

PART I

Approaches to Sustainability and Health Care

CHAPTER 1

Environmental Bioethics

F rom the outset, bioethics was concerned with human ties to the environment. In 1927 Fritz Jahr described bio-ethics (German: *bio-ethik*) as "the assumption of moral obligations not only towards humans, but towards all forms of life."[1] Jahr expressed a compassionate ethics towards animals and plants. And, although human ethics and obligations towards flora and fauna are different, both are premised on an underlying respect. Respect for nonhumans consequently requires a moral justification for their destruction. Jahr promoted a Western, deontological articulation of bioethics, stated as "respect every living being on principle as an end in itself and treat it, if possible, as such!"[2] Almost half a century later, the term "bioethics" appeared in English, with an astonishingly similar meaning.

In 1971 American Van Rensselaer Potter used the term "bioethics" to describe a life ethic for an industrialized society struggling against a precarious ecosystem. For Potter, bioethics was rooted in an intrinsically practical approach to sustainable life, inclusive of the earth and other organisms.[3] In fact, the 1978 *Encyclopedia of Bioethics* was so influenced by Potter's scholarship that it defined bioethics as encompassing the "problems of interference with other

3

living beings . . . and generally everything related to the balance of the ecosystem."[4] Essentially, the environment was an integral part of the original concept of bioethics. Despite having coined the term "bioethics," Van Rensselaer Potter had a subdued impact on the development of the academic discipline, and a second way of defining bioethics appeared in American academia.

In 1979 Tom Beauchamp and James Childress at the Kennedy Institute of Ethics at Georgetown University took a divergent approach to bioethics. They moved away from biotic relational systems and towards the physician-patient relationship. Initially calling it "biomedical ethics," Beauchamp and Childress offered four principles to shape the new field: respect for autonomy, beneficence, non-maleficence, and justice.[5] Widespread consensus about the usefulness of the four principles of biomedical ethics—in tandem with institutional recognition from universities and hospitals—codified normative standards for Western, academic biomedical ethics, today called, simply, "bioethics."

The evolution of the concept of bioethics, which was formerly attentive to nature and ecosystems, into a more technological-individual field gave the appearance that Potter's bioethics was a separate discipline from academic biomedical ethics.[6] Indeed, the current practice of identifying and conflating "bioethics" with "biomedical ethics" has erased the ecological origins of bioethics while simultaneously giving rise to the "new" discipline of environmental bioethics.[7]

Environmental Bioethics

Environmental bioethics is a recognized subdiscipline within environmental ethics and biomedical ethics that permeates university curricula, literature, and academic conferences.[8] It focuses on climate change–related health hazards and the environmental impact of health care as well as nutrition, pest control, natural disasters, and public health.[9] Environmental bioethics has been addressing the intersection of environment, resource use, and human health for decades.[10]

Two basic approaches to environmental bioethics within health care exist, each with a different emphasis. The American model, represented by the Healthier Hospitals Initiative, relies on past initiatives of sustainability and does not seek further avenues for conservation. The British model, represented by the National Health Service's various efforts at carbon reduction, uses government involvement to quantify and limit the carbon emission of the health-care industry. Both are organizational applications of environmental bioethics in health care.

The American Model

Many health-care organizations have adopted a mission of sustainability. They advocate, or participate in, selected environmental initiatives—like recycling water bottles—without attending to the other areas of resource consumption such as emergency transportation, research, or medical offerings. Hospital cafeterias, operating rooms, and buildings are touted as "green" if they move towards organic food or renewable energy. The American model of environmental bioethics operates under the assumption that health care has already become sustainable through prior initiatives and does not need to implement further avenues towards conservation.

The Healthier Hospitals Initiative (HHI) based in Reston, Virginia, is a national campaign to implement sustainable measures in the health-care sector. The HHI is a think tank and information center that provides webinars, networking, strategy sharing, case studies, and written guides to incorporate environmentally sound practices into daily operations of hospitals and health-care systems.[11] The Healthier Hospitals Initiative focuses on six branches of ecological health care: healthier food, leaner energy, less waste, safer chemicals, smarter purchasing, and engaged leadership.[12]

Many of these initiatives mirror other corporations' attempts to "green" their business. For instance, a part of the Healthier Hospitals Initiative "Healthier Food Strategy" includes serving less meat and sugary beverages in hospitals. "Leaner Energy" includes using less air conditioning and replacing halogen lights with light-emitting diode (LED) lights, which are cooler. In other cases, initiatives are connected to issues specific to a clinical setting.

Among the initiatives related to health care, the Healthier Hospitals Initiative addresses the ecological implications of red bag (biohazard) waste, single-use devices and reprocessing, and health-care specific toxins like

Di(2-ethylhexyl)phthalate (DEHP) released from polyvinyl chloride (PVC).[13] Overall, the HHI promotes numerous environmental choices and offers sustainable alternatives within hospitals, thus allowing health-care organizations latitude to choose tailored strategies for their ecological and economic objectives.

Healthier Hospitals Initiative transcends institutional and state lines. Opportunities to participate in the HHI are broad and include sustainable design, purchasing local food, encouraging carpooling, reducing oil expended on bottled water, retrofitting buildings with renewable energy, adding recycling programs, and educating employees on ecology. Each hospital that participates in these initiatives has a unique rationale for doing so and an individualized approach to utilizing the HHI framework.

Partnered hospitals, when they choose to join the Healthier Hospitals Initiative, are then listed on the main HHI website by location and identified by the strategies they have adopted. This type of positive pressure encourages other health-care facilities to join the Healthier Hospitals Initiative.[14] Dignity Health and hospitals in Massachusetts are among the health-care systems that participate in the Healthier Hospitals Initiative. These examples concretize the American model of environmental bioethics in health care.

Dignity Health

Dignity Health was founded by the Catholic order of the Sisters of Mercy in 1986. With over forty hospitals and care centers across California, Arizona, and Nevada, Dignity Health maintains facilities that are both "Catholic" and "non-Catholic."[15] These hospitals follow the common, nondenominational values of "dignity, collaboration, justice, stewardship, and excellence."[16] As part of their commitment to the Healthier Hospitals Initiative, Dignity Health has worked towards eliminating mercury in their hospitals, utilizing plumbing devices that conserve more than 100,000 gallons of water per processor per year, and adopting sustainable-design energy retrofits. They have also been PVC/DEHP-free since 2005.

Additionally, Dignity Health was the first hospital system in California to join the now defunct California Climate Action Registry by voluntarily committing to measure and report all greenhouse gas emissions.[17] They also partnered with Healthcare Without Harm—an international coalition working to reform the environmental- and public-health practices of the health-care industry.[18] These changes have added up to measurable environmental conservation, with one report stating that Dignity Health had reduced their carbon emissions by 244,000 tons and recycled 16.3 million pounds of waste in 2013 alone.

The Dignity Health hospital system also eliminated 1.4 million pounds of plastic and prevented carbon-dioxide emissions equivalent to 42,815 gallons of gasoline by using reusable sharps and pharmaceutical containers.[19] As part of the Healthier Hospitals Initiative, Dignity Health has taken a broad approach to organizational sustainability. A similar environmental trajectory is seen in Massachusetts hospitals that participate in the Healthier Hospitals Initiative.

Massachusetts Hospitals

Massachusetts is a hub of medical invention and innovation. The Boston area, in particular, features prominent and Top Ten–ranked hospitals, as well as numerous teaching and research institutions specializing in health care. Over forty hospitals in Massachusetts—including all ten in the Partners HealthCare system—have joined the Healthier Hospitals Initiative, utilizing an institutional approach to sustainability focused on the quality of food served, the type of energy used to power facilities, and efficient architectural design.[20]

In addition to the usual initiatives of organizational greening, like recycling office paper, many hospitals in Massachusetts have "started rooftop vegetable and herb gardens, and kitchen staff are purchasing more local produce from sustainable farms and seafood from local fishermen."[21] Some hospital menus now read like "farm-to-table" restaurants. In the urbane and wealthy Boston area, this is especially attractive.

Overall, the American model of environmental bioethics in health care relies on past efforts at conservation. Any new concerns about ecology fade into the background as busy hospital administrators focus on running their health-care institutions. In contrast, the British model of environmental bioethics quantifies the carbon emissions of health care and continually reevaluates governmental policy. The National Health Service's carbon-reduction regulations target and identify areas for environmental conservation in the United Kingdom.

The British Model

The United Kingdom's 2008 Climate Change Act set a target to cut greenhouse gas emissions of the entire UK by at least 80 percent of their 1990 levels by 2050 through legally binding carbon budgets.[22] This action required the participation of every sector in the UK, including health care. The United Kingdom has focused on carbon emissions because they are a quantifiable measurement of pollution and other environmental harm. Once calculated, carbon reduction can be implemented, often through carbon capping.

Carbon capping identifies an acceptable amount of carbon emissions, but does not permit emissions beyond that point without ramifications, which are usually economic. Carbon capping is believed to be one of the most effective ways of limiting carbon-dioxide emissions. Despite support from both science and ecology, carbon caps are difficult to implement in some countries—like the United States—and there has been international reluctance to initiate any sort of carbon capping, trading, or binding measures to reduce the amount of carbon emitted worldwide. The United Kingdom, in contradistinction, avoids these obstacles through government oversight and socialized medicine, thus successfully limiting carbon emissions.

Against the backdrop of the 2008 Climate Change Act, the United Kingdom's National Health Service (NHS) drafted *Saving Carbon, Improving Health: NHS Carbon Reduction Strategy for England*, and the National Institute for Health Research (NIHR) *Carbon Reduction Guidelines*. Both evaluate the effects of health care on climate change, following the wisdom that "quantifying the environmental impact of health care is important to determine the potential value of mitigation efforts and to reduce harm associated with health care delivery."[23] The British model of environmental bioethics is structured, uniform, and thorough.

National Health Service Carbon Reduction Strategy

The National Health Service's 2009 *Carbon Reduction Strategy for England* outlines conservationist strategies for medical organizations and clinical research laboratories in the UK.[24] The NHS initiative implemented systematic and aggressive policies to reduce carbon emissions in their health-care system under the twofold aim of carbon reduction and more efficient health care.

National Health Service guidelines for carbon reduction parallel many of the strategies for reducing carbon emissions that were highlighted in the Healthier Hospitals Initiative, such as encouraging carbon-neutral transportation—like walking and biking—eliminating animal-based foods from menus, and reducing water waste. Moreover, several unique proposals, such as reducing the "over-supply" of food to curb obesity and examining pharmaceutical-related carbon emissions, also appear in the guidelines. The NHS recognizes that providing better health care for their citizens will "reduce levels of demands for health services" later, and thus environmental burdens on hospitals and clinics.[25]

In observation of the fact that medical demands put pressure on resources and lead to greater carbon emissions, the West Midlands Cancer Intelligence Unit within the UK pioneered the use of cancer registry data along with Geographical Information Systems (GIS) to calculate the carbon emissions associated with treating breast cancer. They report, "Data comparisons from 1999 and 2004 showed that there has been a 214% increase in total car miles travelled which equates to over 400 tonnes of carbon associated with radiotherapy treatment in the West Midlands. Looking at patient and visitor mileage, and therefore carbon, will prove to be a useful tool in the designing of low carbon patient pathways."[26] The diligence with which the NHS is assessing and modifying activities within their health-care system also appears in the companion to the *Carbon Reduction Strategy*, the National Institute for Health Research *Carbon Reduction Guidelines*.

National Institute for Health Research Carbon Reduction Guidelines

Clinical research is a carbon-emitting activity. Part of the "associated carbon footprint" of the universities, hospitals, and general health-care facilities are the academics in offices entering data, performing a systematic review of relevant research, running clinical trials, or working in the laboratory.[27] Since health-care delivery and health-care research are intricately connected, the United Kingdom's 2008 Climate Change Act required carbon assessments of both institutional branches. The National Institute for Health Research (NIHR) is the branch of the NHS that supports clinical research—case studies, clinical trials, study design, data collection, and trial monitoring. The NIHR *Carbon Reduction Guidelines* were crafted to make researchers aware of the carbon impact of their activities and then offer avenues to modify current clinical research practices and mitigate climate change.

In two notable examples, specific clinical trials in the United Kingdom were assessed for carbon output. The first study analyzed the carbon emissions from a sample of twelve randomized controlled trials funded by the National Institute for Health Research's Health Technology Assessment program. The average emission of each trial was calculated to be 78.4 tons per trial, or 306 kg of CO_2 per participant. The total number is equivalent to the amount of carbon produced in one year by approximately nine people in the United Kingdom.[28] While the average of these twelve clinical trials was provided, it should be noted that each clinical trial had a different carbon footprint depending on variables such as location, enrollment, travel, distribution of materials, technological equipment, and electricity.

In a second example, the National Institute for Health Research analyzed the "Crash" trial, which was a multicenter, international, randomized, controlled trial of the effect of corticosteroids on death and disability in 10,008 adults with head injuries. The Crash trial emitted 630 tons of CO_2 or 63 kg of CO_2 per participant. The total carbon emissions were calculated to be the equivalent of 525 round-trip flights from London to New York for one passenger.[29] These clinical case studies offer tangible examples of quantifying carbon emissions from clinical trials. Following these carbon calculations, the NIHR *Carbon Reduction Guidelines* outlined nineteen key recommendations for carbon reduction for researchers. Three of the recommendations are particularly notable.

First, the NIHR recommends that researchers make full use of current literature by "carrying out systematic reviews of existing evidence before submission of new grant proposals." Ensuring that clinical trials address new questions is essential because each clinical trial has a carbon footprint and "carbon cost occurs whether or not a trial is published."[30] In the absence of significant clinical findings, trial proposals cannot advance and will not be published. Thus, a thorough review proactively reserves carbon emissions for trials that will substantially contribute to scientific advancement.

Second, the NIHR endorses efficient study design, for instance by "answering several questions through one trial" or enrolling patients in more than one study. This was successfully accomplished in the Women's Health Initiative, which had three clinical trials within it: a randomized controlled trial on hormone replacement therapy; an observational study on dietary modification; and a community prevention study on Vitamin D/calcium. Women were allowed to enroll in one, two, or all three trials simultaneously, cutting carbon emissions from gathering separate participants. The National Institute for Health Research estimates that this technique—in this initiative alone—saved 13,550 tons of CO_2 or the equivalent of removing one-and-a-half cars from the roads for an entire year. This also serves patient populations who may benefit from multiple clinical trials.

Third, the NIHR advocates avoiding unnecessary data collection by "measuring outcomes remotely by phone, mail, or the internet whenever possible" and encouraging record linkage (with participant consent) with routine data. Also known as telemedicine, technological innovations in health-care research and delivery support sustainability and competent medical practice. Telemedicine is convenient and timesaving as well.

The National Institute for Health Research *Carbon Reduction Guidelines* were "developed by researchers for researchers . . . to highlight areas where sensible research design can reduce waste without adversely impacting the validity and reliability of research." With the twin goals of carbon management and practical study design, the NIHR believe that they can "reduce their carbon footprint without increasing the administrative burden to researchers."[31] The UK is continually evaluating—and moving towards—environmental sustainability in their National Health Service with governmental support.

Conclusion

Both the American model of environmental bioethics and the British model of environmental bioethics attempt to meet the needs of their particular milieu while also reducing resource use. Numerous stakeholders have attempted to turn the tide of rapid climate change, ballooning carbon emissions, and rampant resource consumption through sustainability in health-care institutions, that is, through environmental bioethics. While organizational practices have emerged that address global concerns and sustainable health care, environmental bioethics has thus far taken a broad approach to reducing the resource use and carbon emissions of health care. And, neither environmental bioethics nor health care are unified as to the best manner to address environmental conservation while simultaneously providing quality health care.

Since environmental bioethics has failed to offer a coherent model of conservation for health care, health-care organizations invested in environmental bioethics lack a uniform dedication to specific ethical principles. Environmental bioethics is largely reactive to climate change. In the American context, environmental bioethics has cobbled together ecological initiatives rather than presenting a seamless garment of sustainable health care from top to bottom. In the British model, environmental bioethics has addressed the environmental impact of the health-care system as an organization without looking within its walls, thus sidestepping environmental accountability at the level of the doctor-patient relationship. As medical resource use continues to increase exponentially, it will not be enough to simply maintain the status quo. What is needed, rather, is a specific, proactive, and systematic approach to sustainable bioethics that all sectors of health care can utilize.

CHAPTER 2

Green Bioethics

ealth care cannot let profligate resource consumption take on "the harmless aspect of the familiar."[1] In this era of the Anthropocene, the complex interface of resource use and industrialized health care requires something much more structured than the current models of environmental bioethics. It requires a biomedical ethic driven by the concerns of environmental ethics. Through four principles, Green Bioethics can bridge remaining gaps in environmental bioethics by addressing sustainability in medical developments, techniques, and procedures. The term Green Bioethics indicates points of contact and divergence with environmental ethics and biomedical ethics, while also extending the environmental bioethics tradition.[2]

Green Bioethics and Environmental Ethics

Green Bioethics is committed to sustainability, as is environmental ethics. In the well-cited *Report of the World Commission on Environment and Development:*

"Our Common Future" also known as the *Bruntland Report*, sustainability strikes a balance between conserving resources for the future and using resources for present needs. On the spectrum of resource use, conservation is between preservation, which safeguards resources without using them, and exploitation, which uses resources without regard for limits. Within this spectrum, resource use can be either sustainable or unsustainable.

Determining the sustainability of resource use depends on a number of factors. Concepts such as IPAT (Impact = Population × Affluence × Technology)[3] and "carrying capacity"[4] have been developed to assess sustainability. Resource use that is temporarily sustainable may be unsustainable long-term. Thus, the idea of longevity is inherent within the concept of sustainability, which "meet[s] the needs of the present without compromising the ability of future generations to meet their own needs."[5] Sustainability—since it is forward-looking—includes notions of justice, conservation, simplicity, and humanism. All of these values are foundational to Green Bioethics, which addresses sustainability as an outgrowth of environmental bioethics.

Green Bioethics and Environmental Bioethics

Both Green Bioethics and environmental bioethics draw on the values of environmental ethics. However, their focus is different. Currently, the most comprehensive model of environmental bioethics—the British National Health Service—examines carbon emissions with the intention of carbon limitation. While carbon-capping models might appear to be a path for Green Bioethics to continue on, this approach is viewed with caution. A "carbon-capping" mentality that supposes there is a set amount of carbon that can be released—but not more—is insufficient for several reasons.

First, the amount of "safe" carbon in the atmosphere—calculated to be 350 parts per million—has already been exceeded.[6] Once carbon is in the atmosphere, natural corrections are long and arduous. High-tech solutions, such as geoengineering, could artificially reduce carbon in the atmosphere, but with unknown ramifications. Geoengineering, moreover, siphons resources and emits carbon through its deployment.[7] Society should not emit more carbon in the futuristic pursuit of carbon reduction. Carbon quotas fictitiously imagine that climate change can be addressed through more consumption.

Second, a carbon-capping mentality has the potential to reduce environmental ethics to the carbon number associated with a given item. Ultimately, this absolves individuals from thoughtful consideration and inner motivation for conservation. Michael S. Northcott forcefully argues, "Markets in carbon are idols that legitimate the continuation of a consumptive industrial economy and the continuing sacrifice of the common goods of a stable climate and a livable earth for future generations."[8] Simply identifying a carbon number and then declaring an item "sustainable" or "not" is morally reductionistic.

Third, national and institutional calculations of carbon do not capture resource use of individuals. Both China and the United States have very high country carbon outputs. However, according to the most recent data from the United Nations, the average per capita carbon emissions were 6.5 metric tons in China, and 2.5 times higher—at 16.8 tons per capita—in the United States.[9] Per capita carbon calculations based on national emissions disguise the vast discrepancies in health-care access and use between individuals in a country. Furthermore, if carbon limits were allocated to each individual within the health-care system, policymakers would encounter issues of justice. Those who are born with health complications would disproportionally suffer from individual carbon medical limits, while those with fewer medical challenges might feel compelled to use the full extent of their carbon limits, regardless of real need.

The aforementioned considerations point to the need for a coherent and sustainable ethic. In many ways, carbon capping is appealing as an approach to sustainability in health care, but it is insufficient to address long-term resource consumption. The carbon emissions of the vast majority of health care have not yet been calculated,[10] and sustainable health care is too urgent a matter to wait for the carbon calculations of every aspect of the health-care industry in each country. Therefore Green Bioethics will offer principles to reduce resource use and examine the sustainability of medical developments, techniques, and procedures in particular.

Green Bioethics and Biomedical Ethics

Green Bioethics takes a circumscribed approach to biomedical ethics by examining the sustainability of medical developments, techniques, and procedures such as artificial life support, genetic enhancement, pharmaceuticals, elective surgery, end-of-life care, in vitro fertilization, and organ transplantation. Medical developments, techniques, and procedures are the focus of Green Bioethics because they are the focus of traditional biomedical ethics. Medical developments, techniques, and procedures are a specific area of health care that is seldom addressed in terms of environmental impact—unlike waste management and electricity use in hospitals.[11] Further, medical developments, techniques, and procedures form a core identity of medical practice and are universal in defining medical care. Medical developments, techniques, and procedures go beyond the hospital and are utilized in numerous health-care settings such as home care, international medical relief, emergency medical care, dental offices, and auxiliary clinics. While Green Bioethics and biomedical ethics have a similar scope of medical developments, techniques, and procedures—the ethical priorities—or principles that are utilized to guide decision making—are distinctive.

Ethical Principles

Among the numerous ethical systems that can be applied to biomedical ethics—narrative ethics, feminist ethics, communitarian ethics, or theological ethics, for instance—a principled approach has been favored by Western biomedical ethics.[12] The 1978 National Commission for the Protection of Human Subjects of Biomedical and Behavioral Research offered respect for persons, beneficence, and justice as three guiding principles in *The Belmont Report: Ethical Principles and Guidelines for the Protection of Human Subjects of Research*.[13]

Tom Beauchamp and James Childress paralleled the principles from the *Belmont Report* when they proposed respect for autonomy, justice, beneficence, and nonmaleficence in *Principles of Biomedical Ethics*. Despite the appeal of consensus-based principles, which articulate "general norms that give considerable room for judgment in many cases,"[14] numerous critiques of principle-based moral theories, or "principlism," have been leveled within and against biomedical ethics.

Daniel Callahan has pointed out that "a longstanding complaint against principlism is that it has never embraced some system of lexical ordering and is thus in a poor position to deal with conflict among the principles."[15] This is primarily a problem of methodology and not content. In the absence of a clear way of applying the four principles of biomedical ethics, individuals may choose the most convenient principle and neglect the others. Without priority, certain principles that are conceptually simpler will be favored for the sake of expediency. Methodology aside, the content of Beauchamp and James Childress's *Principles of Biomedical Ethics* has been called into question as well.

The principles of respect for autonomy and justice have both received ample amounts of attention. A major critique of autonomy is defining it in a way that is hyper-individualistic, thus giving it a privileged place in the Western world, which seems to only grow stronger with each passing year. The other three principles seem to bend towards a particular view of autonomy that makes patient preferences sacrosanct. Yet, this does not need to be the case. Autonomy can take on contours of consideration for others, as seen in Immanuel Kant's articulation of autonomy, for instance. However, the current way of using the principle of autonomy in biomedical ethics is atomistic and overemphasized.

Conversely, justice in biomedical ethics is often relegated to an afterthought and has become "a bit of a 'remainder' concept."[16] The principle of justice in biomedical ethics is simply too broad to be applied to the highly individual sphere of the physician-patient relationship. While the neglected principle of justice was partially addressed in the "four topics" of clinical ethics,[17] it remains in the shadows of biomedical ethics.

Furthermore, justice as a concept is malleable, making it difficult to define and apply. There are many different versions of the concept of justice, ranging from the virtue of justice, to social distributive justice, to power justice, to ecojustice, and intergenerational justice.[18] These versions of justice are able to fit within Beauchamp and Childress's concept of biomedical justice, resulting in inconsistent application of the principle.

Despite objections, principles are appealing as an ethical system because they offer "a set of norms that all morally serious people share as the common morality."[19] Principles build consensus and allow many people to contribute to the formation of morality. Principles are also broad enough to give freedom, but narrow enough to give guidance in particular situations. For this reason, Green Bioethics will use a principle-based system as an analytical framework. But Green Bioethics will not express the current "common morality" of Western, individualistic biomedical ethics, but rather an *uncommon* morality that reflects the origins of Van Rensselaer Potter's bioethics. The complex interface of environmental ethics and biomedical ethics requires something much more creative and interdisciplinary than the current approaches. It requires an approach to the scope of traditional biomedical ethics driven by the concerns of environmental ethics.

Principles of Green Bioethics

Green Bioethics offers a principled environmental framework for evaluating the sustainability of medical developments, techniques, and procedures. Taken in aggregate, the four principles of Green Bioethics articulate a rich and multifaceted approach to resource conservation in health care. They can be summarized by the philosophy that medical developments, techniques, and procedures that reduce resource consumption are environmentally sustainable. Conversely, medical developments, techniques, and procedures that increase resource consumption or expend resources unnecessarily are not sustainable. This echoes the Leopold imperative that "a thing is right when it tends to preserve the integrity, stability, and beauty of the biotic community. It is wrong when it tends otherwise."[20] Hence, the four principles of Green Bioethics promote sustainable medical developments, techniques, and procedures and advocate for the reduction or elimination of ecologically unsustainable medical developments, techniques, and procedures.

The four principles of Green Bioethics integrate the scope of Western biomedical ethics by focusing on specific practices with the values of environmental ethics. The principles of Green Bioethics are formed in the wake of Beauchamp and Childress, and thus do not have the luxury of identifying prima facie ethical principles without context. Rather, Green Bioethics emerges from a particular social and educational location of privilege in the Western world that favors an approach to biomedical ethics heavily weighted towards individuality. Green Bioethics also springs from the effects of this individualism seen in the magnitude of the environmental crisis and environmental ethics that have been developed as a result thereof. Thus, Green Bioethics contends with the collision of autonomy-based biomedical ethics, ecological pressures, and ecological solutions. As such, Green Bioethics offers tenets familiar to environmental ethics in the methodological shape of traditional biomedical ethics.

The principles of Green Bioethics are:

1. Distributive justice: allocate basic medical resources before special-interest access
2. Resource conservation: provide health-care needs before health-care wants
3. Simplicity: reduce dependence on medical interventions
4. Ethical economics: humanistic health care instead of financial profit

It should immediately be apparent that Green Bioethics inverts Beauchamp and Childress's *Principles of Biomedical Ethics* by naming justice first. This upending is an intentional and important distinguishing feature of Green Bioethics.

Methodologically, principle 1 is given lexical priority while the other three principles are intersectional and do not need to be ordered. This offers a clear starting point for ethical guidance and mitigates potential conflict in applying the principles. Because the principles of Green Bioethics are interdisciplinary, their interactions with each other avoid many problems of competition by utilizing a mixture of ecology and biomedical ethics. In addition to a clear methodological rationale for this move, philosophically, there are several reasons for prioritizing justice.

First, justice as the leading principle of Green Bioethics begins where Beauchamp and Childress conclude. This continuity offers an avenue for bioethicists to engage with environmental ethics in familiar terms. In addition to bridging biomedical ethics and environmental ethics, the principle of justice has resonance with the "fourth box" from clinical ethics.[21] Justice remains one of the principles of both biomedical and clinical ethics, thus earning a well-accepted place in health care.

Second, justice is a recognizable principle of environmental ethics. Here, justice assesses needs and resources and then attempts an equal or even distribution of these goods. Justice takes a broad view of the moral community and adjudicates the use of resources both within the current population and between generations. In 1995, Soren Holm observed, "The ethical system propounded by Beauchamp and Childress lacks the necessary resources satisfactorily to handle the ethically complex situations created in the interface between medicine and social justice,"[22] but the prioritization of justice mends this fissure. Thus, it is a useful heuristic for ecologists who wish to engage the health-care system, medicine, and biomedical ethics.

Third, prioritizing justice protects human rights and guards against extreme forms of deep ecology or modern Albigensianism[23] that could promote the involuntary "extermination" of humans for the benefit of the planet.[24] Philosophers have cautioned against the logical extremes of conservation-based health care, which might include compulsory reproductive sterilization or refusing to provide life-sustaining health care.[25] Justice maintains a concern for the impoverished and disadvantaged and acknowledges that resources must be allocated in a way befitting human rights.[26] At the same time, justice refuses to deny clinically indicated medical care simply because humans are environmental liabilities who require resources to live.

Ultimately, justice—and distributive justice in particular—answers the critique that Western biomedical ethics tends towards hyperindividualism, particularly through the principle of respect for autonomy. Distributive justice tempers individualistic tendency by its very nature since it is always situated in a community. This intellectual shift from the individual to the group has deep resonance with ecology, which analyzes matrices of affiliation. Distributive justice acknowledges individual desires and needs without acquiescing to them as the ultimate good. After distributive justice, three further symbiotic principles are offered.

Principle 2 focuses on conservation. Resource conservation arises from an ecological concern for dwindling resources, situated within the reality of health-care needs. Resource conservation recognizes that resources must be used, but that they should be used in a way that all people—current and future—can access them. Jeopardizing the health care of others because of overuse of medical developments, techniques, and procedures will be a failure of medicine, e.g., antibiotic resistance.[27] However, acting with resource conservation will benefit the health of all people. There is no explicit connection between this principle and those offered by Beauchamp and Childress. While the principle of resource conservation leads to positive health-care outcomes—such as quality of life and well-being—it is a departure from the traditional principles of biomedical ethics. Resource conservation is firmly entrenched in ecological ethics.

Principle 3 offers simplicity as a standard for sustainable biomedical ethics. Although the word "simplicity" is most closely identified with the environmental movement, physicians in the health-care system utilize this principle when they act with therapeutic parsimony or diagnostic elegance. Therapeutic parsimony seeks to provide health care in the least invasive and most efficient way possible, and diagnostic elegance relies on Ockham's razor to identify maladies, thinking first of the more obvious causes of illness.[28] Simplicity, in this way, connects to Beauchamp and Childress's principle of non-maleficence since unnecessary medical treatments can harm patients from nosocomial infection and iatrogenic externalities. Each medical development, technique, and procedure has risks, even if they are negligible. In simplicity, unnecessary exposure to health-care harm is minimized.

Principle 4 engages economics. The current American health-care system is situated within a capitalistic machine that is predicated on growing markets. Profitability often determines which medical developments, techniques, and procedures proliferate and which remain dormant. The neglected medical needs of the poor and systemic plunder of natural resources are among the unfortunate consequences of a health-care system that operates as a business. Ethical economics most closely fits with Beauchamp and Childress's principle of beneficence, since it acknowledges that basic health care should be given to all people regardless of ability to pay. Natural resources are directed at greatest clinical benefit, while luxury medical goods are curtailed. Ethical economics represents a new order that effectively provides humanistic health care.

Conclusion

Green Bioethics draws on theory and philosophy from environmental ethics and biomedical ethics, thus creating an interdisciplinary approach to sustainable health-care practices. While four principles of Green Bioethics have been put forth, it should not be assumed that any specific medical development, technique, or procedure must meet all four principles to be "green." These are, after all, *principles*, not rules. On the other hand, adherence to one or two of these principles alone does not guarantee a sustainable medical development, technique, or procedure. The elasticity present in Green Bioethics allows for numerous approaches to sustainability in health care, while also remaining steadfast in the foundational commitment to justice.

PART II

Principles of
Green Bioethics

CHAPTER 3

Distributive Justice

ealth care is caught in an ethical quagmire: utilizing too many resources for specialist medical developments, techniques, and procedures for a few, yet seemingly unable to supply basic medical developments, techniques, and procedures for all. Hospitals in the United States dedicate an enormous amount of resources providing artificial life-support in intensive care units even after there is no hope for recovery. Pharmaceuticals are prescribed for any physical or mental inconvenience, when many times natural solutions would remedy the underlying condition. Patients are offered numerous diagnostic tests over the course of multiple doctor's visits that produce little clinical benefit. On the other side of the globe, the medical situation is vastly different.

Nearly everyone in the developing world is at a disadvantage for accessing basic health care. Broken bones are set by themselves, leading to disfigurement and disability. Lack of maternal care results in poor birth outcomes, intellectual delay, and death in infants. Many children cannot be seen by primary care physicians because of cost or location. Curable diseases become chronic illnesses because medicine is unavailable.

The current arrangement of medical distribution of medical developments, techniques, and procedures is clearly unjust, despite precedence in ethical philosophies to the contrary. Jon Fuller and James Keenan appeal to the priority of justice in health-care ethics, in particular. They aver, "When we attempt to broach the topic of justice not with the individual First World physician, but with the developing world public health official, justice becomes not a remainder concept but rather, the basic conceptual framework for the bioethical discussion."[1] But, in order to secure worldwide justice, several aspects must come together. Health care, doctors, and medical priorities must coalesce to ensure just allocation of medical resources. Technology, policy, and personal responsibility must be aligned to support resource conservation. A minimal standard of medical care must be available and accessible for all people before the medical elite utilizes developments, techniques, and procedures that do not cure, treat, or prevent diseases.

The first principle of Green Bioethics—distributive justice—states that a general allocation of medical resources should precede special-interest access. Distributive justice addresses the tension between resource use and health-care disparities by mitigating gaps between medical access and services. Distributive justice also satisfies the concern of ecologists regarding limited resources, and global medical workers who recognize the need to expand health care worldwide. Since distributive justice reconciles health care and sustainability, it has lexical priority within the principles of Green Bioethics. In health care—as with the planet—most resources are a zero-sum game. Health care must be, first and foremost, just.

Theories of Justice

Theories of justice are often put forth in broad ethical areas, such as politics, philosophy, feminism, and economics.[2] Although there are different expressions of justice, there are three essential similarities that all conceptions share. First, theories of justice are relational. Justice is not an activity or virtue that can be exercised in isolation; it presupposes a social location or community. Justice is impossible without other people.

Second, theories of justice are communal. Theories of justice tend to benefit the common good—however defined. In many cases, relationality is the foundation for the implementation of justice for the benefit of the community. Although theories of justice may include or omit some in the community based on characteristics such as sentience, geography, biological makeup, or species, justice extends beyond one moral agent for the benefit of the many.

Third, concepts of justice are nonexclusive. Justice is usually compatible with an additional ethical framework, making it applicable in the realm of environmental ethics and biomedical ethics. Many theories of justice do not operate autonomously. Rather they account for competing meta-ethical views. For instance, Jessica Pierce and Andrew Jameton note that "environmental sustainability and social justice are mutually reinforcing goals, and both are vital elements of population health."[3] Adequate theories of justice encompass other notions of morality.

Within environmental ethics and biomedical ethics, theories of justice also appear. Environmental justice and biomedical justice have aspects that reflect the relational, communal, and nonexclusive characteristics of justice. To be sure, there are differences between environmental justice and biomedical justice. But in both, justice moves further from practices and ideologies that place a fraction of people squarely within resource access and others devastatingly outside of it. Environmental justice, since it is at the core of environmental ethics, will be considered first.

Environmental Justice

Philosophers, such as the eighteenth-century Immanuel Kant, held that a person could only transgress against a member of the rational, moral community. Since nature, plants, and animals were considered nonrational, they were not members of the moral community and could not experience injustice. However, the owners of these things could have injustice done to their nonrational property. The crime would be against the rational human being, however, and not the property.[4] Kant's thought underpins a modern view that environmental injustice only occurs against humans. Environmental justice, from this perspective, focuses on the disproportionate impact of climate change on disenfranchised human communities, usually in the global South, but also some places in the North. Indeed, the concerns of environmental racism illustrate anthropocentric environmental justice in a profound way.

The concept of environmental racism began in the mid-1980s in the United States when the United Church of Christ (UCC) commissioned a report on racial justice, leading to the 1987 publication *Toxic Wastes and Race in the United States*.[5] The UCC research found that toxic waste sites were closer to poor neighborhoods. Thus, environmental destruction affected poor African Americans more than middle- and upper-class European Americans. While rich people could afford to move away from unhygienic dumpsites that would lead to disease and illness, the impoverished people could not. Furthermore, rich people had the resources to lobby against waste sites in their zones, while poor people suffered from a lack of advocacy.

Anthropocentric expressions of environmental justice—such as environmental racism—are legitimate, but they are ultimately thin views of justice because they assume eco-destruction is deleterious only if and when humans are imperiled. They fail to account for all creatures in the biotic community and ignore human dependence on a thriving ecosystem. Moreover, environmental justice's emphasis on the group does not satisfy the obligation society has to the individual, nor does it balance group entitlements with the needs of the individual.

Biomedical Justice

Biomedical justice within health care narrows the sphere of justice to the individual and is typically anthropocentric, although it may focus on research ethics while using animal subjects. Modern contours of justice in health care have most recognizably been articulated through four principles of biomedical ethics.[6] Tom Beauchamp and James Childress's principles of benevolence, non-maleficence, respect for autonomy, and justice[7] were situated in the wake of gross human-rights violations, such as the ones that came to court at the Nuremburg trials, and a series of unethical medical studies like the Stanford Prison Experiment, the Tuskegee study, and the "Monster Study."[8] The four principles of biomedical ethics were also a response to formally recognizing the inherent value and dignity of persons who should not be instrumentalized, even for the "benefit" of others.

After Beauchamp and Childress's principles of biomedical ethics, modern medicine gravitated towards certain principles, such as benevolence and non-maleficence, as well as noninterference. Autonomy, in particular, has been emphasized as a reaction to the aforementioned medical abominations. Simultaneously, the biomedical principle of justice has been downplayed or relegated to an afterthought. When practiced, justice most commonly focused on physicians avoiding prejudices such as sexism, racism, and classism. Beauchamp and Childress's work in biomedical ethics was a necessary first step in an increasingly complex world marked by advances in medicine, personal liberties, and social change. And, while the principles of biomedical ethics have been reevaluated over time,[9] globalization has profoundly altered modern understandings of justice in health care.

While the four principles of biomedical ethics might have sufficed for well-educated white men who were working in an American context, female, minority, and developing-world biomedical ethicists no longer envision justice from such a standpoint. Significantly, not all communities agree that autonomy should be valued above the common good.[10] A broader vision of bioethics has been articulated in recent decades, one more aligned with Potter's original view of "global bioethics." This dissonance between the global village in which we live and the Georgetown principles leaves a gap in applied biomedical ethics.

Karen Peterson-Iyer thus suggests that health care and medicine should move "beyond the horizon of individual choice to consider the broader well-being of the larger community and its many members and groups."[11] Today, the clustering of medical resources significantly influenced by unrealistic demands of patient autonomy has become a concern in both environmental and biomedical ethics. A more equitable allocation of resources is required to meet the claims of all individual members of society.

Distributive Justice

Distributive justice allocates resources with a view towards fairness, not excess or deficiency. Distributive justice may be enacted in three basic ways: as arithmetical equality, proportional equality, and equality for the common good.[12] Arithmetical equality states that the same number or amount of resources should be given to each person: e.g., one loaf of bread for one person. Proportional equality offers resources relative to need: e.g., one loaf of bread per 100 pounds would translate to a 100-pound man receiving one loaf of bread and a 200-pound man receiving two loaves of bread. Distributive equality for the common good maintains equilibrium between the individual and the group. In this case, bread is allocated towards the people who will be able to use the nutrition to most benefit the common good. In times of war, this might be soldiers; in times of population decline, this might be fertile women. There are benefits and drawbacks in each type of distributive justice.

Arithmetical equality appears to be the most "fair" on the surface because it is plain conceptually and simple to implement, but arithmetical equality does not account for variations in humankind the way proportional equality does. Furthermore, arithmetical equality can result in inequality if some people have more than they need (e.g., an older man may not need an entire loaf of bread), while others have less (e.g., a growing child may need more than one loaf of bread).

Proportional equality meets the reality of human variation, but implementing it is complex. People's needs change over time. A typical female will be a young girl, pregnant woman, nursing mother, and elderly woman in one lifetime. Proportional equality would have to adjust to the ever-changing needs of each individual.

Equality for the common good benefits the most people. In this way the community is as strong as possible. However, equality for the common good may contravene the claims of individuals to a minimally decent life. The old, sick, and disabled who may contribute to the common good in less obvious or quantifiable ways might be disregarded.

Distributive Justice and Health Care

Elementary health care is a fundamental good that must be provided to all people, without exception. At the same time, there appears to be a tension between just distribution of medical developments, techniques, and procedures and environmental sustainability. If everyone in the world consumed medical goods at the rate of the United States, for instance, there would be a drastic increase in resource use. It appears that health care can be either sustainable or just, but not both. "The sheer number of people struggling to live with almost nothing—coupled with the profound constraints of our already stressed ecosystems call into question our ability to achieve both sustainability and justice. We may have to ask which should have primacy," warn environmental bioethicists Jessica Pierce and Andrew Jameton.[13] This view of justice assumes that everyone must have an arithmetically equal and maximal amount of medical resources available to them. When arithmetic equality is assumed, sustainability and just medical distribution compete with each other. However, proportional equality enacts the claims of both distributive justice and sustainability.

From among the three theories of distributive justice, Green Bioethics favors proportional equality because it safeguards human dignity while also recognizing limits to resources. Proportional equality within distributive justice in health care acknowledges that all people have basic medical needs, while some people have extraordinary medical needs. Since there are sharp disparities in wealth, health, and quality of life across the globe, working to mitigate these inequalities is part of distributive justice. Proportional equality thus reconciles environmental justice and biomedical justice by directing resources to general allocation of health care for all individuals and away from special-interest access for the few. Instead of competing for primacy, sustainability and distributive justice are radically dependent on each other, or, as Pierce and Jameton maintain, "mutually reinforcing goals."[14]

In order for health care to be distributed proportionately, many people will need more resources, while some people will require less. Those in the developed world who have their basic health needs met will not require additional resources. Indeed, some people who have accessed basic health care have already been afforded proportional justice. Many more people have violated the demands of justice by using too much. Rich countries that monopolize natural resources must be willing to recognize the entitlements of the global community. The first principle of Green Bioethics—distributive justice—endorses a general allocation of medical resources before special-interest access in health care.

Special-Interest Access and Resource Use

Special-interest access to medical developments, techniques, and procedures can be characterized by four related components. First, special-interest access exceeds health-care needs, which will be defined more thoroughly in the next chapter. More specifically, special-interest access goes beyond general health care for a particular individual. For example, eyewear is a necessity for someone with myopia; it is a choice for someone with perfect vision.

Second, special-interest access occurs when a medical development, technique, or procedure is used that does not cure, treat, or prevent diseases. This can be verified in the course of medical care. Tuberculosis vaccines prevent disease; elective hip replacement does not cure, treat, or prevent disease.

Third, procedures that have a long treatment course and are not available to all people with similar conditions, or are highly technological, experimental, elective, or expensive may indicate special-interest access. For instance, allogeneic bone-marrow transplant is "associated with extremely high mortality; up to 40 percent of patients die from complications of the transplant itself."[15]

Fourth, special-interest access depletes a large amount of intellectual, financial, or communal resources without concomitant medical benefit. There are numerous examples of special-interest access in health care. Non-lifesaving treatments like otoplasty (cosmetic ear surgery), bariatric surgery, Lasik eye surgery, and pharmaceuticals for non-life-threatening allergies are examples of high-level lifestyle, elective, or expensive services offered to those who have disposable income. Since special-interest access in health care is often resource-intensive, curtailing it will reallocate resources to those in need and reduce overall medical use.

Special-interest access may cause very little intrinsic harm; however, when many people use medical developments, techniques, and procedures that are special-interest access, and with greater frequency, there are aggregate harms in the form of medical waste and resource depletion.[16] Furthermore, the carbon impact from these special-interest services contributes to climate change. Climate change–related health hazards compound health needs. Poorer countries that have neither the material resources nor the political resources to defend themselves are disproportionately affected by climate-change health hazards and a lack of medical care.[17]

Health services are usually regressive. The *Lancet* documented that "the distribution of health care expenditures on services other than primary care—mostly higher-level services—[are] skewed toward the best-off."[18] Intellectual, natural, and medical resources are funneled into niche health care that provides non-lifesaving, lifestyle procedures to those in the middle and upper classes. Jessica Pierce and Andrew Jameton rhetorically ask, "Can industrialized countries in the northern hemisphere support their high levels of health care consumption without exploiting or ignoring widespread poverty, environmental degradation, ill health and suffering in poorer regions of the world?"[19] Their question echoes the twofold reality of medical distributive injustice and resource consumption.

The planet has limited resources, and current modes of distribution are both unfair and too resource-intensive. Current offerings of health care must be reprioritized. By way of illustration, the two extremes of the resources required for assisted and natural reproduction highlight the difference between special-interest access and general allocation of resources. On the one hand are women—often with a partner—who plan a medically laborious conception and pregnancy through assisted reproduction, which is frequently tied to further medical resource use such as elective cesarean sections and hospital admissions for their premature infants. On the other end of the medical spectrum are women who often cannot plan or avoid pregnancy and do not have doctors to attend their childbirth or aftercare. They are without neonatal intensive care units and rarely have medical attention even in pregnancy emergencies.

Assisted reproductive technologies (ARTs) are situated within the larger realm of misprioritized elective medical developments, techniques, and procedures in the developed world. Although proportional equality within distributive justice is the standard used in Green Bioethics, fertility treatments are not necessary for health or life; they do not cure, treat, or prevent disease; they are low-success and high-tech; and they are not available to all people. Therefore it will not be argued that ARTs should be more widely available. Rather, ARTs will be examined in terms of basic medical resources and special-interest access. The fertility industry—not infertility prevention or parenthood—is the subject of the following ethical assessment.

Special-Interest Access, Resource Use, and Assisted Reproductive Technologies

In the developed world, women have a wide variety of ob-gyns to choose for supervision during natural pregnancy care, trained personnel to attend births, and a much lower maternal mortality rate than the developing world. Couples and individuals have many options available to them in terms of how they would like their children to be conceived, gestated, and delivered. Many of these possibilities are used in conjunction with the assisted reproductive technologies industry—a high-tech, resource-intensive industry accessible only to a special-interest population.

Assisted reproductive technologies are an encompassing term for various reproductive developments, techniques, and procedures in fetus-making that include fertility treatments, in vitro fertilization (IVF), artificial insemination (AI), surrogacy, pre-implantation genetic diagnosis (PGD) for selection or elimination of disabilities, sex selection, "designer babies," and "savior siblings." ARTs also include recent advances that splice deficient mitochondria from an egg, replace it with healthy mitochondria from a donor that does not carry the targeted genetic anomaly, fertilize the egg, and then implant the zygote into the womb, resulting in three biological parents. ARTs temporarily replace—not restore—infertility.[20] Assisted reproductive technologies are not simply a medical procedure for couples who are infertile. Rather, they are used by fertile people, including fertile same-sex and opposite couples, and fertile gay and straight single people.

There are over 464 reporting fertility clinics in the United States alone.[21] Fertility clinics and treatments grossed $16.1 billion dollars in 2013.[22] Clinicians are attracted to this branch of elective treatments because of the large salaries. Ian Craft, director of the United Kingdom's Humana fertility unit, "revealed that in the late 1980's some practitioners were making over £500,000 annually from their infertility practice."[23] Today, fertility doctors are still drawn to these lucrative positions that offer hundreds of thousands of dollars a year.

Marketing strategies, media coverage, and growing availability increase interest and use of elective assisted reproductive technologies among those in the capitalistic developed world. "Boutique" in vitro fertilization clinics offer clients the choice to "shop" for the "perfect" child using a catalog of donated gametes and desirable surrogates. Then, in conjunction with a geneticist, a woman is impregnated in a clinic from zygotes created in laboratories.

Once pregnant, the woman can schedule an appointment to deliver the child or children via cesarean section. Cesarean sections are a cosmetic choice for many, but not all, women. If it is only done to retain the elasticity of the vagina, in order to stay "honeymoon fresh," then it is clearly a symptom of medical consumerism.[24] Delivery by cesarean is often medically unnecessary. Furthermore, cesareans are correlated with higher postpartum antibiotic treatment, severe maternal morbidity and mortality, and worse outcomes for the infants.[25] One study found that "small, premature infants born by cesarean section are at 30 percent higher risk for serious breathing problems than those delivered vaginally."[26] Despite this, it remains a popular—and resource-intensive—choice among U.S. women.

The United States has one of the highest percentages of elective cesarean sections in the developed world—at nearly a third of all deliveries—despite the World Health Organization's warning that overuse of unnecessary cesareans is a "barrier to universal [medical] coverage."[27] In the developed world, after the planned, elective, and vaginally aesthetic cesarean section, women emerge from the hospital having utilized a vast panoply of medical resources, doctors, and services. The multiplicity of options and additions in the fertility business reflect the consumerist use of ARTs.

Fertility treatments are voluntary and could be avoided, but governmental provision of reproductive technologies for virtually any adult in the UK[28] and coverage of assisted reproductive technologies by some health-care insurance companies in the United States[29] are expanding, rather than attenuating, the use of these procedures in the developed world. They are special-interest procedures, which are also marketed as adjuncts to—although separate from—clinically necessary treatments such as chemotherapy for cancer.

Special-Interest Access, Resource Use, and Oncofertility

Oncofertility is concerned with possible fertility "loss" due to cancer treatments. It is primarily available in the developed world, where people have a very good chance of surviving some cancers. Oncofertility options for men have been developed;[30] however, since the male role in assisted reproduction is limited, and there are no harmful medical side effects of natural sperm harvesting (i.e., masturbation),[31] oncofertility options for men are not as resource-intensive as oncofertility options for women and will not be addressed here.

Oncofertility possibilities for women include oocyte cryopreservation, embryo cryopreservation, and ovarian tissue cryopreservation. These are not exclusively offered to cancer patients and may be used by people without cancer. However, they highlight the penetration of special-interest access of ARTs into a necessary medical branch (oncology), thus confusing and blending the two.

Oocyte cryopreservation is the storage of female gametes in a very cold storage facility for an indefinite amount of time. It can be used for women undergoing chemotherapy. It has also been suggested that people who voluntarily seek contraceptive sterilization and sex-reassignment surgery—even after informed consent has indicated that fertility will be terminated—should be offered fertility preservation instead of performing these services after procreative choices are complete.[32] Additionally, some scholars have supported making egg freezing available to all women for nonmedical reasons.[33] The American-based company Facebook offers cryopreservation of women's eggs as part of their health-care plan, and the United States military is also contemplating cryopreservation for female soldiers. Sperm and egg banks use cryopreservation for donor genetic material, and couples use cryopreservation for their unused embryos while undergoing assisted reproductive technologies.

Embryo cryopreservation is the storage of a fertilized egg. Embryos are stored by couples undergoing ARTs to be used at a later time or disposed of.

Embryo cryopreservation can be used by cancer patients with the intention of conception after successful treatment, by cancer patients where there is little hope of successful cancer treatment and their partner wishes to reproduce posthumously following the death of the cancer patient,[34] and by singles and couples who are not ill. As of 2007 there were "over 400,000 frozen embryos in cryopreservation storage facilities in the United States."[35] Embryos can be stored indefinitely.

Embryo and oocyte cryopreservation can be used by women who are of reproductive age; ovarian tissue cryopreservation is available for females who are too young to be fertile as well as women who are reproductively mature. In 2012, the youngest girl to forcibly undergo ovarian tissue removal was two years old.[36] Minors—especially toddlers—cannot consent to medical treatments. It is ethically dubious to compel children to have invasive non-lifesaving treatments, particularly if resources are limited. Furthermore, fertility cannot be *preserved* in children. Since children are only potentially fertile, they can have procedures done that can aid them in medical reproduction later in life. Notably, a child who is rendered infertile by cancer treatment does not actually lose anything that she ever had and cannot be made fertile with these treatments.

Oncofertility sets into motion a series of medical procedures centered on reproductive technologies. They cannot increase the chance of surviving cancer. If a person is cured, the former cancer patient must use assisted reproductive technologies for a chance of biological reproduction, unless ovarian tissue has been frozen. If ovarian tissue has been frozen, it can be retransplanted into the woman's own body after cancer treatments.[37] Oncofertility and assisted reproductive technologies always use resources. However, they are not always successful in their intended outcome of live birth.

The American Pregnancy Association records that, in the United States, the live birth rate for one round of IVF per age group is as follows: 41–43 percent for women under age thirty-five; 33–36 percent for women ages thirty-five to thirty-seven; 23–27 percent for women ages thirty-eight to forty; and 13–18 percent for women ages over forty.[38] The low success rates of IVF are comparable to pregnancy rates by ovarian tissue transplantation, which offers an average 30 percent chance of pregnancy and birth.[39]

The retrieval, preservation, storage, and treatments following fertility preservation use substantial medical, intellectual, and natural resources. From an environmental perspective, oncofertility and cryopreservation are an unmitigated resource drain without concomitant medical benefit, since the inability to become pregnant does not threaten life or physical well-being. While certainly medicine should attempt to prevent, treat, and cure cancer in all parts of the world, oncofertility is part of the reproductive technology business, not cancer care. This is compounded by the fact that ARTs—linked to, or independent of oncofertility—use additional special-interest medical resources before and after pregnancy.

Special-Interest Access and Resource Use before and after ARTs

Assisted reproductive technologies are not stand-alone procedures that merely absorb resources one time. Rather, they are one component of a multistep process that requires additional medical resources. These begin with preconception fertility treatments and culminate in intensive care for the children born from ARTs. They use resources that do not benefit—and often harm—women and children. They are special-interest access and often require more medical developments, techniques, and procedures to care for the effects of use.

Before an egg can be fertilized, fertility drugs are given to a woman to artificially stimulate multiple egg production in order to facilitate simultaneous multiple egg removal. However, these drugs can lead to ovarian hyperstimulation syndrome (OHSS). The *American Journal of Reproductive Immunology* states that OHSS "is one of the most important complications of ovarian stimulation with severe morbidity and is still a threat to every patient undergoing ovulation induction."[40] Venous thrombotic events are a further outcome of OHSS, placing women at risk for death.[41] Moreover, fertility drugs are known carcinogens and have been linked to increases in cancer in women following ART births.[42] Women who use health care for ARTs must often be treated for these and other conditions that are a direct result of their conception choice. The assisted-reproductive-technology pregnancy is laden with special-interest medical resource use after birth as well.

Women seeking ARTs often choose to become pregnant with twins and require medical attention that is less frequently needed with singleton births.[43] These are generally not spontaneous, and thus avoidable, episodes of twinning. The *European Journal of Obstetrics & Gynecology and Reproductive Biology* indicates that "the incidence of twins after ART born at <32 weeks increased 27-fold from 1987 to 2010 and has not reduced from its peak incidence over the last decade."[44]

Twin pregnancies are medically undesirable. One report indicated that 8 percent of babies born in the United States were low birth weight and 11.4 percent were premature.[45] When used with ARTs, gestating twins reflect the choice to transfer multiple embryos with the intention of multiple implantations. These are usually for social reasons like "wanting to make up for lost time" or the convenience of limiting pregnancies. Yet, when comparing singleton and twin pregnancies, one multicountry survey found that potentially life-threatening conditions were 2.14 times higher; maternal near miss was 3.03 times higher; severe maternal outcomes were 3.19 times higher; and maternal deaths were 3.97 times higher in twin pregnancies.[46]

Twins or higher-order pregnancies often result in premature and low birth-weight infants, which require extra medical attention and resources. These conditions are frequently addressed while they are in the neonatal intensive care units (NICUs), or in the months after. Studies demonstrate that twins conceived using in vitro fertilization have longer birth admission to NICUs in their first year of life. These infants are 60 percent more likely to be admitted to a neonatal intensive care unit and have higher incidence of hospital admission later in life.[47] In these cases, the NICU becomes one of the final stages of chosen, premeditated clinical conception. The journal *Human Reproduction* concludes, "Estimations of the cost of an ART twin delivery should include the extra 4 days on average spent in hospital at birth, the almost 4-fold increased risk of admission to a NICU and the increased risk of hospital admission in the first three years of life."[48] In addition to the serious medical complications related to prematurity and being born at a low birth weight, any time spent in a neonatal intensive care unit carries "significant" iatrogenic disease and medical-error risks.[49]

Those who access medical goods in industrialized countries have been accustomed to a system that serves their demands for medical developments, techniques, and procedures. Bioethicists and policymakers cannot ignore the fundamental lack of distributive justice coupled with massive resource use in the developed world. Assisted reproductive technologies are resource-intensive and ethically controversial. Utilization of ARTs are increasing, and with them issues of use and access. Thus they will remain features of health-care ethics for years to come and will need to be reassessed as a growing sense of environmental responsibility, global justice, and poverty-reduction coalesces. Pregnancies resulting from assisted reproductive technologies are instructive for examining sustainability in the health-care industry, but should not be thought of as the only application for the first principle of Green Bioethics. Distributive justice is also an imperative for a general allocation of medical resources.

General Allocation of Medical Resources and Resource Use

Women in developed countries often have control over their fertility and can typically prevent multiple consecutive pregnancies that may threaten life. They often give birth in clinical settings attended by trained professionals, with the best medical technology available. These births are in contrast to unattended, often rural deliveries by women in many parts of the developing world who suffer the physical, emotional, and social consequences of maternal complications. Martha Nussbaum poignantly notes that in many geographical locations, women "have not chosen the lives they lead, since frequently they have no conception, or a deficient conception, of alternatives, and a confined list of possibilities."[50] This extends into the realm of conception, procreation, and child-rearing.

At the same time, women's health is often drastically affected by numerous unplanned pregnancies, which relegate their lives to medical hardship. Maternal care and maternal doctors are examples of general allocation of medical resources that would support distributive justice with minimal resource use. Extending maternal care to all women has little resource impact, significant clinical benefits for women, and compounded advantages for the newly born infants.

General Allocation of Medical Resources and Maternal Doctors

Since doctors provide medical development, techniques, and procedures, their placement is a necessary prerequisite of health care. If it were the case that all locations had a proportional number of doctors for citizens, the first principle of Green Bioethics could directly address medical development, techniques, and procedures. However, before that is possible, just doctor placement and medical developments, techniques, and procedures must remain tied together.

Physicians trained in maternal care are needed worldwide. The World Health Organization indicates that a ratio of 23 doctors, nurses, and midwives per 10,000 people is the minimum number necessary to deliver essential maternal and child health services.[51] However, in forty-nine priority countries, forty-four have less than the minimum number, with over 70 percent having less than 10 health-care practitioners per 10,000 people.[52] The lack of maternal-care doctors providing medical developments, techniques, and procedures is devastating for women in impoverished circumstances.

For millennia women have been giving birth without hospitals or doctors. At the same time, even preindustrial labors were assisted by community professionals like doulas or midwives. These birth attendants are absent in many births in the modern world. While the developed world has nearly 100 percent of births attended by skilled health personnel, parts of rural Africa experience fewer than 30 percent of births attended by skilled health personnel. Worldwide, this translates to millions of live births without adequate care.[53]

Medical doctors, nurses, emergency medical technicians, midwives, and health-care paraprofessionals can all be enlisted in the service of providing maternal care. A physician does not need to assist the birth of a child; a midwife or a trained labor-and-delivery nurse could mitigate maternal mortality and morbidity through her skills. Without trained professionals, women die unnecessarily in labor or suffer life-altering complications that could have been prevented. The global community must make strides to provide maternal doctors to care for millions of women without.

General Allocation of Medical Resources and Maternal Care

Maternal care is a universal health concern. Yet, basic forms of maternal care are not equally distributed among women. The World Health Organization (WHO) reports that factors that prevent women from receiving or seeking care during pregnancy and childbirth include "poverty, distance, lack of information, inadequate services and cultural practices."[54]

The matters of maternal health care and distributive justice are complicated by the biological and systemic connection between fertility, sex, sexuality, and societal power. Pregnancy taboos and misunderstanding about sex and reproduction thwart proper maternal care, as well as global apathy. Indeed, "Across the globe women's wellbeing is threatened by early or late childbearing and substandard care during pregnancy and childbirth. In some developing regions infertility is as much as three times higher than in developed countries due to inadequate healthcare, unsafe abortions, undiagnosed or untreated pelvic infections, and botched delivery."[55]

Medical issues in maternal care are compound by a lack of antenatal care. Medically complicated, unattended pregnancies often have long-lasting repercussions. In the developing world, women who suffer from maternity-related fistulas, and cannot access reparative surgery, are ostracized.[56]

In sum, current maternal-care pattern allocation violates the principle of distributive justice. It is one aspect of a complex system of self-interest, economic pressure, governmental indifference, and policy inaction.[57] Pregnancy-related complications are problematic worldwide, yet significant research and resources are focused on special-interest access to technological reproduction instead of general allocation of basic maternal care.[58]

ARTs absorb medical resources and contribute to the destruction of the environment. This is a salient ethical factor in the distinction between the pregnancy that could have been prevented—or safely brought to term without extra medical resources—and conception, pregnancy, delivery, and aftercare that are chosen despite scarce resources and multiple ethical issues. A general allocation of maternal care, before special-interest access to assisted reproductive technologies, would save lives, reduce the environmental impact of health care, and support just distribution.

Distributive Justice and Sustainable Health Care

To apply the principle of distributive justice, health care must simultaneously expand general allocation of medical resources to those in the developing world and reduce special-interest access in the developed world. It would be reckless to increase the consumptive lifestyle of the developing world until it matches that of the developed. That would result in exceeding current resource use.[59] And, even if health care brought poor people into a higher standard of health care while continuing to offer elective treatments to the developed world, it is likely the rich would never be satisfied with their own standard of living. They would perpetually want to be "ahead" and would not be content with equal access to health care, always pursuing special-interest access. The Aristotelian concept of *pleonexia*, or the insatiable desire for more, is an apt label for this vice. Distributive justice is the most fitting principle to address this ethical dilemma.

The expanding fields of telemedicine and teleclinics are related initiatives that have redistributed health care to those without medical resources. Telemedicine gives impoverished patients access to health care in remote parts of the world, while teleclinics reduce resource use through digital medical consultations. They highlight distributive justice in practice and exemplify a general allocation of medical resources.

Telemedicine

Digital medicine, or "telemedicine," is defined as "the use of electronic information to communicate technologies to provide and support healthcare when distance separates the participants."[60] The proliferation of smartphones and the internet, in tandem with the portability of telemedicine, place it in the realm of general allocation of resources, not special-interest access.

Telemedicine encompasses both services and delivery mechanisms. The American Telemedicine Association, which was established in 1993, describes the services of telemedicine as primary care and specialist referral services, remote patient monitoring, consumer medical and health information, medical education among telemedicine services, and networked programs that link tertiary-care hospitals and clinics with outlying clinics. Delivery mechanisms of telemedicine that utilize medical developments, techniques, and procedures include point-to-point connections using high-speed networks, monitoring centers for in-home care, and Web-based, electronic-health patient service sites as platforms.[61] Telemedicine also includes electronic intensive care unit (e-ICU) centers, where "care of patients during the stabilization process, evaluation for the appropriateness of transfer, and the prolonged provision of critical care during delays in transport" can be utilized.[62]

Telemedicine has the potential to conserve resources in health care in two important ways. First, use of telemedical services moderates travel to and from clinics. This reduces resource use since patients and physicians do not have to use transportation that relies on fossil fuels, such as car, bus, or tram. The National Institute for Health Research recommends clinician and patient participating in telemedicine through "measuring outcomes remotely by phone, mail, or the internet whenever possible" and encouraging record linkage—with participant consent—to routine data.[63] Additionally, resources will be saved since a third location—the clinic—will not need to provide electricity, water, and temperature control for doctor's visits. Rather, these doctor's visits will occur within the homes of the physician and patient. Because of the enormous benefits to patients, ease of use, and potential for environmental savings, the National Health Service recommends that "all NHS organisations should make training and equipment available that promote tele, video and web-conferencing."[64]

Second, and simultaneously, telemedical platforms conserve paper through electronic medical records (EMR).[65] EMR platforms prevent record printouts, streamlining health care and aiding in better patient outcomes. Electronic medical records not only prevent deforestation, they also reduce resources by not depending on machinery to process and make paper. In 2014, EMR Modernizing Medicine, in Boca Raton, Florida, created the electronic medical assistant (EMA).[66] EMA is just one among many electronic medical records that demonstrate sustainability.

In addition to reducing resources, electronic medical records exhibit distributive justice by making doctors and health-care providers who supply medical developments, techniques, and procedures more accessible. For instance, EMRs eliminate obstacles to continuity of health care by providing a unified system to trace records of migrants and refugees, who often forgo health care while resettling.

Andrew Thorniley believes that telemedicine "will allow instant access to all patient records, ongoing treatment and other consultations . . . *The future is digital*."[67] Telemedicine, inclusive of e-ICU centers and electronic medical records, reduce resource use from travel, limit pharmaceutical waste by monitoring adherence, and save thousands of pounds of paper from record printouts. And, if telemedicine were fueled by renewable sources, it could be totally carbon-neutral. Telemedicine assists in the provision of medical developments, techniques, and procedures by laying the infrastructure for such delivery. In addition to the many environmental benefits of telemedicine, the ability to reach those in developing and underdeveloped areas—particularly through teleclinics—indicates distributive justice.

Teleclinics as a General Allocation of Resources

Teleclinics are virtual health-care centers that "provide an opportunity for standardization and equity in provision of healthcare, both within individual countries and across regions and continents."[68] Teleclinics mitigate physician shortages by remotely connecting patients to doctors. They also employ health-care assistants to provide basic medical care under the supervision of a doctor, who is linked in virtually. Advocates of telemedicine warn that it "cannot be [a] substitute for physicians in developing countries where resources are scarce and public health problems are in plenty . . . however, it can supplement the current health scenario."[69] Teleclinics in India are a case study of distributive justice in health care.

Aparajita Dasgupta and Soumya Deb report that during 2005–2008 the Indian Space Research Organisation telemedicine network "expanded to connect 45 remote and rural hospitals and 15 super-specialty hospitals." And, with periods of overlap during 2006–2008, the pilot project in Karnataka provided more than ten thousand tele-consultations.[70] India, in particular, is a dramatic beneficiary of teleclinics due to its urgent medical requirements, poverty, and large population. In some Indian cities, women have particularly benefited from teleclinics.

World Health Partners, started in 2008, is an Indian-based organization that works primarily in reproductive health care for women.[71] Within three years, World Health Partners "established 116 telemedicine clinics providing health services to 1,293 villages with an estimated population in excess of 6 million people."[72] Patients travel to teleclinics to utilize the services and receive basic medical care. Video conferencing and telemedical programs read the women's blood pressure, temperature, heart rate, and respiratory rate. Electrocardiogram tests are also given. After this data is obtained, patients are counseled on contraception. Over 288,000 "couple years" of contraception were delivered during three years.[73] Considering the substantial need for spacing pregnancies in the developing world, the teleclinics of India have undoubtedly saved many lives through offering contraception, and improved the medical quality of life of millions of women.

Telemedicine will be a feature of sustainable health care in the twenty-first century. New developments include Aetna's collaboration with Teladoc, a "U.S. board-certified licensed [doctor] in your state" that patients can contact twenty-four hours a day, an artificial intelligence program that acts as a computerized therapist, and machine-learning that analyzes a new mother's activity and posts on Facebook and Twitter to predict postpartum depression.[74]

Telemedicine and Distributive Justice

Although telemedicine is an integral part of sustainable health-care delivery, telemedical services are still inaccessible to many people in the two-thirds world for a number of reasons. Obstacles to implementing telemedical clinics include the financial cost of modern buildings, consistent electricity, and compatible electronics to run teleclinics.[75] These logistical challenges reveal a tension in the Western exportation of solutions to global problems.

Just as "leapfrog" technology was suggested for developing countries to move beyond rudimentary energy production from burning wood and coal, thus bypassing carbon-intensive energy production,[76] so too is telemedicine impractical for many countries. It is not enough simply to present telemedicine to the developing world; rather, the basic structures necessary for human life and health care must be in place as well.

Gaps in medical need and available options for health-care distribution reiterate the inequalities in global distributive justice. The very poor require food and water, sanitation facilities, and basic protection from violence prior to any offering of sustainable and just medical developments, techniques, and procedures through telemedicine. A comprehensive package of health care would need to include electricity, buildings, running water, and staff if medical teleclinics were accepted as an application of distributive justice.

Nonetheless, telemedicine has the potential to move towards a general allocation of medical resources by connecting patients to health care worldwide. Those outside of urban areas, in very remote places, and in underserved countries could have contact with doctors through telemedicine and access to medical developments, techniques, and procedures that they would not normally have access to. When the infrastructure to deploy telemedicine is brought to developing countries, telemedicine can effectively meet the requirements of distributive justice and reduce resource use.

Conclusion

The health-care industry is not faced with a shortage of medical supplies for basic needs that must be rationed. Instead, the current health-care system simply does not prioritize care for all. Substantial amounts of resources are concentrated in medical developments, techniques, and procedures that have no clinical benefit. The current Western health-care industry displays a deep unwillingness to allocate resources equitably, and an inability to grasp the seriousness of environmental destruction in the face of unnecessary resource use. The marginalized, the poor, and women in particular suffer because of a lack of basic health care. Bioethicists and policymakers must make every effort to find viable avenues to enlarge the supply of medical developments, techniques, and procedures to those in the two-thirds world as a matter of distributive justice. Indeed, justice demands that a general allocation of medical resources is provided to all people before special-interest access.

It is time that the connections between excess medical use and environmental destruction, global justice and sustainability, are made in biomedical ethics and clinical practice. "Given the enormity of the health risks associated with environmental health burdens, and given the attention that bioethics has devoted historically to questions of distributive justice, it is a glaring omission that bioethics . . . [has] devoted so little attention to questions of environmental justice to date."[77] One person cannot consume without affecting another. Sustainability and justice are compatible. If basic medical resources are allocated before special-interest access, the benefits to the planet and its inhabitants will be innumerable. If neglected, both people and planet are imperiled.

CHAPTER 4

Resource Conservation

onsumer demand drives the provision of health-care wants. This absorbs natural resources unnecessarily and places undue pressure on the environment. At the same time, health-care needs go unmet. Many people are unable to access basic medical care, suffer from preventable diseases, and die tragic, early deaths. Environmental bioethicists Jessica Pierce and Andrew Jameton advise, "Along with society at large, health care should accept a responsibility to meet current needs in ways modest and clean enough to be sustainable for centuries."[1]

It is undeniable that the earth has a limited amount of resources. Consequently, it is necessary to curtail some of the environmental "spending" in industrialized health care. The tension between limited natural resources and the imperative to expand health care to all people—present and future—appears dilemmatic. But increasing health care at the expense of resource use is a false dichotomy. The second principle of Green Bioethics states that environmental conservation can occur, in part, when health-care needs are provided before health-care wants.

Health-care delivery must confront the reality of resource limitation and act accordingly. Prioritization of health-care needs will not conflict with environmental conservation if health-care wants are limited. Quite simply, a medical industry that provides for the health-care needs of all people instead of providing health-care needs *and* wants will reduce the amount of material and natural resources used.[2] Determining which medical developments, techniques, and procedures can be classified as health-care needs and which developments, techniques, and procedures are health-care wants must be rooted in a prior understanding of basic human needs and wants. The goal of these distinctions is to lay the foundation for sustainable health care, which can provide health-care needs for all.

Human Wants and Needs

Ethicists must first be able to identify some of the shades of basic human needs and basic human wants as a prerequisite to saying anything meaningful about health-care needs and wants. The demarcation between basic human needs and basic human wants is not precise, but some generalizations can be made. These generalizations clarify contested concepts. The objective is not to provide a definitive list of basic needs and wants. Such a list would be unlikely to reach unanimity, even within nations or health-care systems. Rather, the task is to identify the shades of basic needs and wants in order to guide ethicists and policymakers through this conceptual miasma.

Basic human wants are superfluous, above what is necessary for somatic and social human life. These may also be called desires or hobbies. Some desires may speak to human need, such as a desire for fast food when one, in fact, needs a balanced meal. Some hobbies fulfill a human emotional need for belonging or connection with higher pursuits. Desires and hobbies are not inherently unethical or resource-intensive, but they can easily become material pursuits. As human desires move into the marketplace and drive the economy, they often drain natural resources unnecessarily.

The commodification of desires through extensive marketing campaigns and mainstreaming may cause wants to be seen as needs. For instance, automobiles and entertainment-based electronics are purchased to make modern life easier, but at their core, they are wants. Ecological conservation requires clarity about basic human wants when resources are limited.

Basic human wants occupy a large conceptual space above basic human needs. That is, what constitutes a basic human want can be debated and may change from era to era, or country to country. Some basic wants are relative to availability; a person could not want a yacht if they have never seen one. Other wants are driven by peer group or social location. Even as basic wants proliferate, there remains a consistent sense of basic human needs.

Many basic human needs are collectively recognized across various cultures, religions, and geographical locations. Academic discussions on basic human needs must not become obfuscated by colloquialisms. For instance, a person might say, "I need coffee in the morning." But, empirically, coffee is not a need. While a caffeine dependency may exist, somatic function does not depend on caffeine intake, even if one performs optimally with a certain amount of caffeine in the blood stream. On the other hand, it would be accurate to say, "I need to eat." Macro- and micronutrients are basic human needs. What, precisely, a person needs to eat to survive is a subset of nutritional need. A useful propaedeutic in the quest for collectively recognized basic human needs is Martha Nussbaum's work on capabilities. Her capabilities are not needs, but articulate the human expression of life. The human expression of life points to the shared human condition out of which basic needs arise.

Martha Nussbaum's Human Expression of Life

Martha Nussbaum's capabilities approach was developed from the work of economist Amartya Sen[3] and "aim[s] to be as universal as possible, and its guiding intuition, in fact, directs it to cross religious, cultural, and metaphysical gulfs. For . . . we do recognize others as human across many divisions of time and place."[4] Identifying the essence of human life can direct us to basic human needs and then health-care needs. Nussbaum articulates a human expression of life through a two-tiered approach.

Nussbaum observes that humanity is expressed first by a "shape to the human form of life" and second by "basic human functional capabilities." The first level details the aspects of human life that make it *human*, recognizing the limits and possibilities of embodied anthropological existence. Nussbaum's shape of the human form of life includes mortality, the human body, capacity for pleasure and pain, cognitive capability (perceiving, imagining, thinking), early infant development, practical reason, affiliation with other human beings, relatedness to other species and to nature, humor and play, and separateness.[5]

This list is quite basic and has little moral value associated with it, being more descriptive than prescriptive. Since the focus is on the distinctiveness of human beings, the primary function is to catalog what makes one human identify another as human. It should be noted that not all of Nussbaum's "shapes" are necessary for biological existence, but each of the "shapes" is a necessary precursor to the second level of basic human functional capabilities.

The second level of human expression builds on the first level and provides some implications of humanness. Nussbaum's basic human functional capabilities are:

- Being able to live to the end of a complete human life, as far as possible; not dying prematurely, or before one's life is so reduced as to be not worth living.
- Being able to have good health; to be adequately nourished; to have adequate shelter; having opportunities for sexual satisfaction; being able to move from place to place.
- Being able to avoid unnecessary and non-beneficial pain and to have pleasurable experiences.
- Being able to use the five senses; being able to imagine, to think, and to reason.
- Being able to have attachments to things and persons outside ourselves; to love those who love and care for us, to grieve at their absence, in general, to love, grieve, to feel longing and gratitude.
- Being able to form a conception of the good and to engage in critical reflection about the planning of one's own life.
- Being able to live for and with others, to recognize and show concern for other human beings, to engage in various forms of familial and social interaction.

- Being able to live with concern for and in relation to animals, plants, and the world of nature.
- Being able to laugh, to play, to enjoy recreational activities.
- Being able to live one's own life and nobody else's; being able to live one's own life in one's very own surroundings and context.[6]

Nussbaum's description of human existence is universal, but not complete. These capabilities are, in the words of Nussbaum, intentionally "thick" and "vague,"[7] thus allowing for permutations across time. Their expressions vary by culture. Since these forms of human life are foundational—and in Nussbaum's writing, widespread—they are sufficient for grounding the concept of basic human needs. Capabilities might also identify human wants by contradistinction.

While some of Nussbaum's basic human functional capabilities, based on the shape of human life, are relevant to health care—such as being able to live to the end of a complete human life, being able to have good health, being able to avoid unnecessary and non-beneficial pain, and being able to use the five senses—others are not. Thus, Nussbaum's capabilities are offered as a discursive tool for ethicists working from an environmental framework, moving towards identification of health-care needs. To be sure, health care has its own canon for defining the intersection of humanness and medical practice, based on the goals of medicine.

Goals of Medicine

The traditional "goals of medicine," described by Joseph H. Howell and William Frederick Sale in 2000, are a standard for the health-care industry.[8] These goals of medicine guide the purpose and practice of health-care professionals. The goals of medicine provide an explanation for the purpose of medicine, and thus health care, but do not provide a list of health-care needs or health-care wants. That is, the goals of medicine in practice address a health-care need without specifying the means to that goal.

These goals hinge on a regressive set of terms, but still have cachet. In order to understand the concept of "medicine," a constellation of ideas that revolve around notions of "health," "disease," "illness," and "sickness" must also be defined. Although the goals of medicine are intimately tied to notoriously ambiguous concepts, it is still possible to define the goals—or core values—of medicine, given a general conception of "health."

In 1946, the World Health Organization (WHO) defined health as "complete physical, social and mental well-being, and not merely the absence of disease or infirmity."[9] This definition of health endures as a reference point for those working in health care, yet Howell and Sale astutely note that health care alone cannot fully realize or address the vast compendium of criteria included in the WHO definition.

While the goals of medicine aim at providing physical health to all people, certain aspects of health are contingent on factors outside of health care's jurisdiction, namely, those related to social well-being. Therefore, Howell and Sale strategically focus on health care and propose four basic goals of medicine, which are interlocking and "have a greater or lesser importance under different circumstances."[10] Howell and Sale's goals of medicine are:

- The prevention of disease and injury, and the promotion and maintenance of health.
- The relief of pain and suffering caused by maladies.
- The cure of those with a malady, and the care of those who cannot be cured.
- The avoidance of premature death and the pursuit of a peaceful death.

These four goals of medicine can, in part, determine if a medical development, technique, or procedure will meet a health-care need.

The first goal—the prevention of disease and injury, and the promotion and maintenance of health—is addressed in public health and preventative care. Howell and Sale indicate that smoking-cessation programs would fall under this heading. Many other public-health measures—like minimizing the use of pesticides, mandating seatbelt use, and efforts to curtail obesity—fit within this goal as well. The manner in which these goals might direct health-care needs is flexible.

The second goal is the relief of pain and suffering caused by maladies. This pain and suffering is primarily physical and attended to by pain-relieving drugs, palliative care, and therapeutic surgery. Howell and Sale also include mental anguish and mental disease and disorders within this goal of medicine.[11] This latter aspect of health care requires a particular cautiousness and sensitivity. The history of abusing the concept of "mental illness," for instance, by labeling communists, homosexuals, and suffragettes as mentally ill, and subsequent medical torture—such as electric shock, lobotomies, and gavaging—is a disgraceful legacy of the medical system. Thus, professionals must approach mental disorders, even those with a presumed physical basis, with the utmost caution and discernment.

The third goal of medicine is the cure of those with a malady and care for those who cannot be cured. Under this goal, medicine seeks to "return a patient to a state of normal wellbeing and function."[12] In absence of a full recovery, medicine takes on a different shade, accepting limitations and focusing on care. The "caring" aspect of this goal is especially important because it can encompass end-of-life issues, rehabilitation, support of those at the terminus of chronic illnesses, and geriatrics. This adds a distinctly human aspect to the goals of medicine and reiterates the universality of the human need for interpersonal relationships.

The fourth and final goal of medicine is the avoidance of premature death and the pursuit of a peaceful death. Doctors and patients alike must accept the limitations of medicine and self. The most obvious limitation of medicine is that it can only postpone—but never evade—death. Howell and Sale summarize the ultimate goal of medicine in this statement: "The primary duty of medicine and health care systems is to help the young become old, and then . . . to help those that are old to live out the remainder of their lives in dignity and comfort."[13] Hospitals are not meant to be pedagogues of existentialism and cannot be expected to maintain life support extensively as families grapple with a terminal prognosis. When death is inevitable, medical treatments must transition to comfort care.

The fourth goal of medicine coalesces with the other three, as care related to death may involve components of public health to prevent premature death (from the first goal), provide palliative care at the end of life (from the second goal), and include a sensitivity and care for those who are dying (from the third goal). The goals of medicine retrace accepted ground in the health-care industry, making the concept of health-care needs clearer.

Health-Care Needs

There is consensus that health-care needs exist and that they can be identified in some foundational way. The medical industry expresses the idea of health-care needs through academic and institutional publications, health-insurance coverage, medical language, and prioritization by humanitarian medical organizations. Medical procedures that necessitate proportionate, ordinary, or usual care, as defined in medical literature, could be considered health-care needs. This might include pain management, setting broken bones, and cautious use of antibiotics. Health insurance generally covers these treatments, especially in the developed world. A further way to determine needs comes from clinical nomenclature.

The terms "medically indicated" or "medically appropriate" can identify particular medical developments, techniques, and procedures as a possible health-care need. Insulin shots are a medically indicated need, but only for a diabetic person. In contrast, a non-diabetic person could not claim insulin shots are medically indicated. There is no clinical basis, and much harm and waste that would occur if a non-diabetic started injecting insulin. Border wars occur over what is medically indicated and what is not. These often surface when health-care developments that were intended to treat a medically indicated condition are later offered as elective, enhancing, or lifestyle procedures. For instance, hormonal contraception is effective in safeguarding women from a life-threatening pregnancy and is used as a medically indicated need when taken for this purpose. However, when hormonal contraception is taken so that one may avoid a menstrual period during vacation, there is not a correlative medical indication.[14]

Furthermore, medical developments, techniques, and procedures that are provided by humanitarian organizations can also indicate health-care need. When resources are scarce, these not-for-profit organizations make concerted efforts to triage essential medical care. For example, Doctors Without Borders provides contraception, vaccinations, antiretroviral treatments, and other forms of medical relief.[15]

Since human existence requires use of sensory capacities, working vital organs, a functioning muscular-skeletal system, and mobility, human health-care needs include prevention of maladies, the treatment and care of diseases, the prevention of a premature death, and a painless death. The aforementioned health-care needs are meant as a floor and not a ceiling. For those in health care, the "ontological notion of dignity marks a threshold, a kind of respect and care beneath which the treatment of any human being should never fall."[16] Certainly, these health-care needs can be elaborated upon. And hopefully they would continue to expand as health-care needs become available to all people worldwide.

In order to utilize the second principle of Green Bioethics more effectively, further analysis must be done on medical developments, techniques, and procedures that are not at first glance health-care needs, keeping in mind that the essence of Green Bioethics is more important than debates about the minutiae of its implementation.

Identifying Health-Care Need or Health-Care Want through Two Paradigms

Undoubtedly, drawing the line between health-care needs and wants will be a complex exercise. There will not always be a clear demarcation. Some objections to medical prioritization based on health-care need are made in a spirit of selfishness and anxiety for material luxuries: that is, the apprehension around limiting or refusing opportunities for medical consumerism. Other objections to Green Bioethics might come from the "merchants of doubt": climate change deniers.[17] Both protestations are irrational. There is almost no way that logic can change the mind of someone dogmatically set on consumption despite scientific data, so these arguments will not be addressed. However, other objections to medical prioritization come from those who foresee logistical issues with the dichotomy of health-care need and health-care want themselves. This is a valid concern.

An appropriate resistance to speaking in terms of "health-care need" or "health-care want" comes from the recognition that there are situations where medical developments, techniques, or procedures do not clearly fit into the category of health-care need or health-care want. For example, prosthetics and wheelchairs may not seem to be a health-care need, because not every person requires them. However, since mobility is often a prerequisite for obtaining other human needs like food and water, basic wheelchairs and simple prosthetics for the mobility-impaired are health-care needs.

Contextualization plays into perceptions of health-care needs and wants as well. What may be extraordinary technology in developing countries (e.g., respirators for critically ill newborns) is generally not extraordinary in developed countries. Economic availability, health-insurance coverage or out-of-pocket expense, consumer desire, age, sex, and ability can factor into unclear cases of health-care need. Unclear cases of health-care need—such as breast implants after mastectomy (not reconstruction of pectoral muscles), synthetic growth hormones for diminutive children, pharmaceuticals for "attention deficit disorder," and amniocentesis—require attention because improper categorization endangers access to health-care needs while also risking the provision of unnecessary and resource-draining health-care wants.

To counter this humanitarian and ecological minefield, two paradigms can address situations where medical developments, techniques, or procedures are not immediately apparent as a health-care need or a health-care want. The first paradigm originates from biomedical ethics and examines the contested demarcation between medical function and medical enhancement. Within biomedical ethics, function correlates with health-care need, while enhancement correlates with health-care want. The second paradigm emerges from ecology and examines the distinction between quality of life and standard of living. Under the ecological paradigm, quality of life indicates health-care need, but standard of living indicates a health-care want.

Green Bioethics thus draws on biomedical and environmental ethics from two different but complementary points making the very amorphous concepts of "health-care needs" and "health-care wants" more concrete. These paradigms recognize medical developments, techniques, and procedures as a continuum.

Function vs. Enhancement: The Contribution of Biomedical Ethics

The first way to differentiate between health-care need and health-care want utilizes the biomedical model of therapy or function, and enhancement. As a foil to enhancement, which is generally regarded as medically unnecessary, bioethicists have worked with the category of "therapy," which is generally deemed to be within the goals of medicine. Andrea Vicini notes,

> Frequently, in bioethical discourse, the discussions on technological incorporation have been formulated in terms of the distinction between therapy and enhancement. As this dyad goes, at least in most cases, therapies should not raise ethical concerns, because they aim at promoting healing and, as such, human flourishing. Enhancement, on the contrary, requires more careful discernment.[18]

The discussion surrounding these terms traces back several decades. In 1998, Erik Parens proposed the distinction between enhancement and therapy.[19] Later, in 1999, Gerald P. McKenny favored the use of the terms "therapeutic" and "nontherapeutic" over "enhancement" and "therapy."[20] By 2014, Michael Hauskeller contended that the distinction between enhancement and therapy had been dissolved because the public perception is that enhancement *is* a type of therapy.[21] Biomedical ethics, however, retains the distinction.

Enhancement itself is difficult to define, but it generally denotes a baseline functioning, which is then added to. Bioethical concerns about enhancing medical developments, techniques, or procedures often originate from an anthropological concern for maintaining a human identity.[22] Other ethical considerations are related to justice, access, and of course, potential for abuse. For Green Bioethics, the ethical content of therapy or enhancement is located in resource use.

While enhancement has commonly been paired with therapy, Green Bioethics differentiates between enhancement and *function* for several reasons. Function is more precise, and therefore less subject to misinterpretation, than "therapy." Function has a specific, achievable goal with a clear terminus for medical intervention—that is, bringing a person to a level necessary to obtain human needs, in a way appropriate to age and other limitations of the individual.[23] In contrast, therapy may involve a prolonged period of treatment without end. Function resonates with Howell and Sale's third goal of medicine: "return[ing] a patient to a state of normal wellbeing and function,"[24] thus retracing established ground in health-care ethics. Therapy often has a faddish or trivial connotation to it (e.g., hydrotherapy, aromatherapy). Although the term function cannot escape ableist critiques, neither can therapy.[25] Thus, function will be used.

Medical function does not have to be perfect, but as Daniel Callahan writes, "decent."[26] While even the term "decent" is open to interpretation, it is pithy enough to convey a general sense of use, without digressing into long qualifications at each turn. In determining the difference between enhancement and function, several considerations should be made.

Function or enhancement must account for the unique abilities of each individual. The goals of medicine are not meant to homogenize everyone into one mold of "ability," but rather work with what Jean-Jacques Rousseau considers "natural inequalities."[27] Each person is a mixture of endowed, natural (i.e., biological, genetic) characteristics, capabilities, and developed strengths. A young girl might have an aptitude for speed, but she can also become a fast runner through training and determination. Of course, people are also limited in various ways, and certain physical abilities will be out of range for some. A paraplegic could not become a fast runner through training at this time in medical history, although she might be an Olympic wheelchair-race athlete. A functional or enhancing medical development, technique, or procedure must be assessed relative to each person's baseline.

Function or enhancement is also relative to people in other stages of life. Age-related conditions demand special attention. Middle adulthood modifies the body. "Low testosterone," "infertility," "slow metabolism," and "menopause" are defined as physical deficiencies in some health-care milieus, but these are normal parts of embodied human experience for all people of a certain age. Whether a medical development, technique, or procedure provides function or enhancement depends on the life stage of the individual. One illustration is hormone replacement therapy for menopausal women.

Hormone replacement therapy (HRT) is a treatment given to pre-, peri-, or postmenopausal women to "restore" the loss of hormones like estrogen and progesterone. HRT is carcinogenic and linked to an increase in the risk of stroke and venous thromboembolic events.[28] Hormone replacement therapy "has little if any benefit" according to the *Cochrane Heart Group* journal,[29] yet in the developed world women are often prescribed synthetic hormones to alleviate some of the effects of menopause—like sweating, decreased libido, and bone density loss. HRT does not eliminate menopause; it only masks the symptoms of a normal part of aging.

At the same time, hormone replacement therapy highlights health disparities worldwide. Women who die prematurely are not offered HRT for menopause because they do not live long enough to feel the effects of natural hormonal fluctuations. It is largely an offering for women who have secured an extended lifespan and desire medical intervention for a basic inconvenience. Overall, hormone replacement therapy does not meet a health-care need; it is a medical enhancement, at a particular stage of life, in developed countries. While there are cultural, personal, and social components attached to some age-related degenerations, possible alternatives can also form the ethical assessment of function or enhancement.[30]

Elective hip and joint replacement are now considered a "routine" procedure for people—usually older, white, middle-upper-class Americans with health insurance—who begin to have joint pain.[31] Elective hip and joint replacement do attend to the human need for mobility. Yet, joint pain is oftentimes related to lifestyle and can be prevented or reversed. Joint pain is more prevalent in the overweight and obese who place undue pressure on their joints through their excess weight.[32] Weight loss can alleviate the underlying issue. Noninvasive physical therapies are also an option. The elective nature of some hip and joint replacements, in addition to the presence of alternatives, indicates enhancement. It should also be remembered that not all elective procedures are successful, and many have negative medical externalities.

Incommensurate medical risk to clinical benefit can indicate enhancement. The *British Medical Journal* reports, "Major elective surgery contributes to intensive care occupancy, with a significant mortality rate."[33] Iatrogenic and nosological infections, side effects, and damage to the person—both physical and psychological—can accompany surgery, drugs, and manipulations of the body. While additional medical problems may be present in procedures that address function, the risks vis-à-vis exposure to other diseases and mortality for enhancement are disproportionate to the medical benefit. There is no clinical benefit for enhancement, since by definition it is not a *medical* concern.[34]

Biomedical ethics uses the paradigm of function and enhancement to identify health-care need or health-care want. Medical developments, techniques, and procedures that provide function are typically health-care needs. In contrast, medical developments, techniques, and procedures that enhance beyond what is necessary for human function, as stated above, are generally wants. In order to conserve resources, medical developments, techniques, and procedures that are health-care needs aimed at function should be prioritized before health-care wants that are enhancement.

Quality of Life vs. Standard of Living: The Contribution of Ecology

A second way to distinguish between health-care need and health-care want comes from ecology. Environmental ethics utilizes various lexicons to arbitrate the necessary from the desired. Conservationist Paul Ehrlich, for instance, differentiates between "quality of life" and "standard of living."[35] To be sure, there is debate about what can be included in these two categories, just as there is in the division between function and enhancement. Yet, many distinctions remain in both environmental and medical philosophy.

In ecology, quality of life pertains to nonmaterial aspects of life like love, play, laughter, kinship, society, and nature. In ecology, there are emphases on connections to ecosystems—flora, fauna, and humans—as well as appreciation for the natural world. Quality of life in society prioritizes virtue, the common good, and harmony. One's own perception of her quality of life is subjective because people have different values. Quality of life is something all people seek, though they may do so in destructive ways (e.g., drug use) or constructive ways (e.g., volunteering). Despite variations, overarching similarities in the human shape of life unify a concept of quality of life that is recognizable across individuals and cultures.

In health care, quality of life describes how well one experiences embodied existence. Cognition, function, freedom from pain, mobility, interpersonal interactions, and longevity are included here. Medical discussions about quality of life typically focus on end-of-life issues and aid-in-dying legislation, but medical quality of life is also a consideration in prospects for recovery and return to baseline function. Because each person has a different baseline, medical quality of life is highly individualized. Even so, health care provides a standard for medical quality of life through an assessment of quality-adjusted life years (QALY), which are a combination of health states and their duration.[36] Quality of life, in both ecology and health care, is contrasted with consumerism, or medical consumerism, which fits into "standard of living."

In ecology, standard of living is concerned with the systematic pursuit of material goods. Standard of living cannot measure non-quantifiable aspects of human existence such as how happy one is, how much love they can experience, or how connected they are to society. Standard of living can, however, identify how large one's house is, how new one's car is, or how much money one has in the stock market. Standard of living is often in direct opposition to conservation because it is premised on consumption. Certainly, some forms of standard of living have very little resource impact—such as retirement accounts and academic degrees. But, overall, standard of living is situated in a classed society that connects personal value with disposable income, and individual worth with purchasing power.

In health care, standard of living tends to be associated with economics, acquisitiveness, and luxury goods. Any sort of "aesthetic medicine" is included here, although these procedures vary in terms of resource use. There is a proliferation of "cosmetic dentistry," "aesthetic dermatology," "aesthetic vein centers," elective surgery, body sculpting, spas, and body "boot camps" for medical consumers to choose from. Of course, society puts immense pressure on people—especially women—to maintain their outward appearance. Environmental ethicists and bioethicists can bring this to light, while at the same time being clear that aesthetic medicine increases one's standard of living and not medical quality of life.

Medical standard of living is often recognizable through direct business to consumer marketing, which drives medical consumerism. While these developments, techniques, and procedures utilize the advancements of scientific technology and are offered under the aegis of health care, they do not fulfill a health-care need and are miscategorized as medicine. For instance, one person might request a prescription for Botox "anti-wrinkle" cream after seeing an advertisement in a magazine. Another person might want to utilize chemical compounds to mimic skin tanning after seeing an article on artificial pigmentation-darkening cream.[37] Or, a retiree might seek shoulder surgery so he can play golf in his gated community because a commercial presented the idea. In other cases, medical developments, techniques, and procedures are not covered by health insurance because they contribute to standard of living, not quality of life. Such is the case with gestational surrogacy.

Gestational surrogacy is used by heterosexual couples, same-sex female couples, and single women if a woman cannot or does not want to carry a pregnancy to term. It is also used by same-sex male couples and single men who do not have a uterus with which to gestate a fetus. Gestational surrogacy is part of the elective assisted-reproductive-technology industry and relies on in vitro fertilization (IVF) or artificial insemination (AI), depending on the desires and sex configuration of the people or person contracting the surrogate. While IVF and AI are considered by some to be medical procedures to remedy infertility,[38] surrogates are not infertile, and this procedure does not meet a health-care need of the person using the procedure: the surrogate mother.

One may object that certain medical developments, techniques, and procedures—such as organ donation—are done on behalf of another person, with no clinical benefit to the self. But these cases differ from surrogacy in that the outcome provides a health-care need for the other person (i.e., functioning organs), whereas surrogacy does not meet a health-care need for the person or people using the surrogate. Prevention of pregnancy can be a health-care need, but surrogacy only shuffles this risk onto another person instead of avoiding it altogether. Surrogate motherhood does not address the health-care needs of the surrogate since pregnancy can only cause physical harm.

Indeed, the 2004 landmark *People of the State of Michigan v. Jason William Cathey* states, in part, "By necessity, a woman's body suffers 'physical damage' when carrying a child through delivery . . . Apart from the nontrivial discomfort of being pregnant (morning sickness, fatigue, edema, back pain, weight gain, etc.), giving birth is intensely painful . . . These types of physical manifestations to a woman's body during pregnancy and delivery clearly fall within the definition of 'bodily injury,' for the manifestations can and do cause damage to the body." Using a surrogate offers standard of living, as the person who has hired the surrogate retains full freedom from gestation for nine months, avoids the demands of breastfeeding, and of course, does not put her body through a physically altering and burdensome condition like pregnancy.

To be sure, some consider children to provide quality of life since parents avoid the social stigma of being childless/childfree.[39] Children also provide social capital, which can make life easier in a pro-natalist world.[40] But, this is social quality of life, not medical quality of life. Children are not a health-care need and can be obtained by other means that do not require medical intervention, such as adoption or foster parenting. Health-care policies do not cover a surrogate gestational carrier because, among other reasons, using a surrogate mother does not cure, treat, or prevent disease. If gestational surrogacy were to be considered a health-care need related to medical quality of life, resources would be used unnecessarily.

Prioritizing health-care needs can become part of the ethos of conservation in health care, but it will take collaboration and the efforts of numerous ecologists, medical professionals, ethicists, and humanitarian advocates positing "an ethically appropriate balance between immediate individual health needs and sustainability."[41] It is not the degrees between "quality of life" and "standard of living" that are the concern of Green Bioethics, but rather the large gap between health-care need and health-care want. As ecologists and bioethicists determine ways to conserve medical resources—without reducing the distribution of health-care needs—prioritizing medical developments, techniques, and procedures that support quality of life will lead to sustainable health-care delivery.

Resource Conservation and Sustainable Health Care

Health-insurance policies determine which medical developments, techniques, and procedures will be covered at no cost to the policy holder; which will have a co-pay or deductible; and which will not be covered at all. Health insurance exists, in part, to ensure that health-care needs are not cost-prohibitive. These plans necessarily make choices about which treatments are health-care needs and which are health-care wants. In 2011, when the United States Affordable Care Act (ACA) passed, certain medical developments, techniques, and procedures were included as part of the minimum benefits guarantee with the intention that they would be provided to those in need.

Provision of Contraception and Health-Care Needs

Section 42 U.S.C. 300gg-13 of the 2011 Affordable Care Act mandates that employers "offer insurance coverage of certain 'essential' health benefits, including coverage of 'preventative' services." Among the preventative services were nineteen forms of contraception that the Food and Drug Administration (FDA) has approved, including the male condom, female condom, diaphragm with spermicide, sponge with spermicide, cervical cap with spermicide, spermicide alone, oral contraceptives (combined pill), oral contraceptives (progestin-only), oral contraceptives (extended/continuous use), patch, vaginal contraceptive ring, shot/injection, Plan B, Plan B one-step and Next Choice (levonorgestrel), Ella (ulipristal acetate), copper intrauterine device (IUD); IUD with progestin, implantable rod, sterilization surgery for men (vasectomy), sterilization surgery for women (trans-abdominal surgical sterilization), sterilization implant for women (transcervical surgical sterilization implant).[42] The ACA did not include non-FDA forms of birth control, such as the rhythm method or abortion. The contraceptive mandate was investigated for violating the religious freedom of certain denominations for moral reasons, not clinical ones.[43] Indeed, "the majority [report] assumed that the government has a compelling interest to promote free access to contraceptive agents."[44] Providing contraception under insurance is a significant move towards prioritization of health-care needs for several medical reasons.

First, we live in a society where women cannot simply choose to abstain from sexual intercourse if they want to. Women, in particular, are vulnerable to rape, sexual abuse, and sexual exploitation. And even in a relationship where sexual activity is implied, such as marriage, women often lack the power to negotiate for contraceptive use. Third-party provision of contraception can mitigate some of the outcomes of sexual violence.

Second, there is a gap between women's desire to delay or avoid having children and the availability of contraception. If all women wanting to avoid pregnancy used modern family-planning methods, unintended pregnancies would decline by 71 percent.[45] There are also persistent barriers to women accessing the contraception of their choice.[46] Yet, contraception can prevent pregnancy-related mortality and morbidity and life-threatening sexually transmitted diseases (STDs).

Approximately eight hundred women worldwide die from preventable causes related to pregnancy and childbirth each day.[47] The World Health Organization estimates that 80 percent of all maternal deaths originate from severe bleeding, infections, high blood pressure during pregnancy, and unsafe abortions.[48] Young women ages fifteen to nineteen are at risk for maternal mortality at an average of 52 deaths per 100,000 births. In parts of sub-Saharan Africa, the numbers reach into the hundreds.[49] Furthermore, the United Nations reports that in 2010 there was an average of 210 maternal deaths per 100,000 live births, with parts of Africa experiencing 500 deaths per 100,000 live births.[50] Women in the industrialized world are at risk for maternal mortality as well.

A woman's lifetime risk of maternal death is 1 in 3,800 in developed countries.[51] Women in the United States, for instance, are in the second quintile of countries worldwide for maternal mortality. The United States ranks 136 out of 183 nations for maternal mortality, with other industrialized countries like the UK, Switzerland, and Belgium surpassing the U.S. for maternal care. Still-developing countries like Estonia, Qatar, and the UAE fare better in protecting pregnant women against maternal death than the United States.[52] Part of this is attributable to systemic racism in the health-care industry and health-care disparities. In the United States, black women die in higher numbers than white women during childbirth, or in the months after.[53] Contraception can prevent these health hazards.

Third, when the potential for pregnancy-related harm is combined with the threat of disease transmission via sexual intercourse, the stakes are even higher for fatalities. Unprotected intercourse results in death for many women worldwide due to transmission of human papillomavirus (HPV), which can cause cancer,[54] or human immunodeficiency virus (HIV) transmission. Barrier forms of contraception are highly successful in avoiding these medical risks. Outcomes of conception and pregnancy-related complications are responsible for the second highest percentage of disability-adjusted life years (DALYs)—a measurement of how disease and injury reduce lifespan—for women in all countries, with the disproportionate burden falling on low-income countries.[55]

Because a lack of contraception can cause physical pain, infirmity, disease, disability, and mortality, the prevention of pregnancy and STDs is a health-care need for all women. Contraception is in line with the goals of medicine and should be widely accessible for all women—nationally and internationally.[56] Contraception is a health need as much as a vaccine or antibiotic. The Affordable Care Act correctly included it in basic health care and serves as one model of prioritization of health-care need. Although resource conservation was not the primary driving factor in determining minimum essential benefits, insurance plans can prioritize health-care needs and make choices about which medical developments, techniques, and procedures to cover. Offering health-care needs before health-care wants will direct health care towards sustainability.

Conclusion

Hospitals and health-care facilities have historically focused on meeting health-care needs, but this is no longer the case. Today, clinics overshoot the traditional goals of medicine and have become environmentally unsustainable. Anthropogenic climate change significantly threatens the welfare of the populace. Health care is partially to blame. David Crippen portends, "Our inability or unwillingness to say no is an immovable object . . . There is a strong likelihood that American health care will burst in totally unpredictable ways as the irresistible force meets the immovable object. We will all be observers of this process in our lifetime."[57] The impact of virtually limitless access to medical developments, techniques, and procedures that provide health-care wants is one way the health-care industry has refused to say "no," much to the detriment of the planet and its inhabitants. The environmental crisis is an opportunity to reconfigure health care and inscribe sustainability within health care.

Sustainable health care requires an integrated concept of environmental limitations, the goals of medicine, and health-care needs and wants. Medical developments, techniques, and procedures that support health-care needs use resources, but with tremendous clinical value. Determinations of health-care need can rely on the biomedical model of function versus enhancement, and the ecological model of quality of life versus standard of living. Once health-care needs are provided for all people, and conservation is embedded in health-care delivery, then medical developments, techniques, and procedures in between health-care need and health-care want could be allocated in ever-widening circles. The crux of resource conservation is not that health-care wants should never be provided for, but rather they must be secondary to health-care needs.

CHAPTER 5

Simplicity

Medical developments, techniques, and procedures are a cause of growth for the health-care industry and nation. Students from other countries flock to the United States to be trained in surgical robotics, pharmacy science, and in utero surgeries on fetuses. Cutting-edge, advanced, and improved techniques allow for varied and individual approaches to treatments, but numerous medical interventions into the human condition drain natural resources. The very way in which the Western world approaches health care must be reconceptualized to account for the limits of the planet. It is time to step back from the medical industrial complex[1] and question the interventionist model, which currently underpins health care in America and many other industrialized countries.

The third principle of Green Bioethics has its closest affinity with environmental ethics. Simplicity—reduce dependence on medical interventions—takes a twofold approach to sustainability in health care. First, simplicity proposes that medical conditions and diseases should be prevented whenever possible. Second, simplicity suggests a gradational approach to medical interventions if prevention is not possible. Health care must consider the limits of our shared resources. Simplicity should not be seen as a reduction of options, but rather an enlarging of a clean and sustainable health-care industry.

Simplicity in Environmental Philosophy and Biomedical Ethics

Simplicity qua reduction of consumption is a prominent feature of environmental philosophy and a growing area of interest in biomedical ethics. In ecology, the "voluntary simplicity" (VS) model has been promoted as a means to conservation for over two decades.[2] In biomedical ethics, simplicity is demonstrated in diagnostic elegance, therapeutic parsimony, and antibiotic stewardship. Both environmental and biomedical streams of thought are premised on the acknowledgment of limited resources and the desire to do more with less. Simplicity is a personalist approach that ties the holistic interests of human beings with an ethic of conservation.

Simplicity in Environmental Philosophy

The Voluntary Simplicity movement in environmental philosophy is premised on the idea that less is more. Voluntary simplicity has two aspects. First, VS decreases consumption of material goods through "downshifting." Second, VS increases noncommercial experiences, values, and relationships. The aggregate result is less resource use, greater happiness, and increased personal satisfaction. Voluntary simplicity is distinct from involuntary poverty, which is undesirable for multiple reasons such as limited upward mobility, fewer social and economic opportunities, and social prejudices. Voluntary simplicity is enticing to those in the developed world who are reconsidering their consumption and reevaluating the toll that the modern world takes on one's inner being.

The first part of VS, "downshifting," eliminates unnecessary consumer goods like duplicate cars, redundant electronics, and complicated material lifestyles. Downshifting is significantly associated with sustainable household practices,[3] as there is less money to expend on unnecessary goods and services. Those who downshift find themselves with less purchase power, but often with more time, energy, and fulfillment. For instance, voluntary simplicity favors meals cooked at home with ingredients selected by the preparers, as opposed to going out to eat at a restaurant.

The double-dividend hypothesis states that there may be two results from one action.[4] Such is often the case in voluntary simplicity. In the aforementioned example of downshifting, people save resources that would be expended on transportation to and from the restaurant, while also increasing their satisfaction in the kinship ties that are developed around a home-cooked meal. This double dividend leads directly to the second aspect of simplicity—an expansion of nonmaterial enjoyment.

The second part of VS endorses nonmaterial satisfaction and might include enjoying a walk with a significant other, reveling in the changing autumnal leaves, volunteering one's time to a nonprofit group, or contemplative practice. Here, simplicity focuses on social activities, connection with nature, community involvement, and creativity. There are many sustainable options for interpersonal connection, such as reading a book and discussing it with a group of friends over tea. Voluntary simplicity accentuates the intangible aspects of life rather than the resources expended on the event. In ecological philosophy, simplicity becomes an approach to life that is founded on gratitude, wonder, and peace. Voluntary simplicity works towards reflective living that fully engages the moment and recognizes that money cannot buy happiness.

Environmental philosophy advocates simplicity through decreased consumption, and enlarges one's internal life with intangible goods. Voluntary simplicity is primarily attractive to wealthy, overworked, and overtired citizens of the developed world who are generally warm, well-fed, and seeking a better quality of life. In many ways, VS is a leisure-class phenomenon that is impossible for those who must work long, exhausting hours and have limited options for sustainable purchasing. While VS is largely driven by individuals, the industrialized health-care system has also been advocating simplicity, with growing momentum.[5]

Simplicity in Biomedical Ethics

Biomedical ethics uses the framework of simplicity without naming it as such.[6] Nonetheless, the characteristics of simplicity—such as limitation, sparing usage, and reflection before use—are a part of diagnostic elegance, therapeutic parsimony, and antibiotic stewardship. A first illustration of simplicity in biomedical ethics is the concept of diagnostic elegance. In the words of Daniel Sulmasy and Beverly Moy, "Diagnostic elegance means using only those tests that are necessary to find out how best to help the patient, not to satisfy curiosity or to feed one's own (or the patient's) love of technology. Important steps in the recognition of low-value and wasteful care in medicine . . . identify wasteful practices and develop claims-based measures of low-value services."[7] The rationale behind diagnostic elegance is primarily financial; unnecessary diagnostic tests waste hospital money. Furthermore, there is little clinical value to redundant tests once a diagnosis can be established with a high degree of certainty. Simplicity in diagnostic elegance emerges as fewer tests are offered and undergone. Once a diagnosis is established, biomedical ethics encourages therapeutic parsimony.

Like diagnostic elegance, therapeutic parsimony attempts to use the least amount of medical intervention to treat an underlying disease or condition. Edward Pellegrino and David Thomasma note that the principles of diagnostic elegance and therapeutic parsimony derive from Ockham's razor.[8] Ockham's razor, developed by William Ockham in the early fourteenth century A.D., is a philosophical approach to problem solving, which states *pluralitas non est ponenda*, meaning the simplest approach is preferable to a more complicated approach. In biomedical ethics, "therapeutic parsimony means using only as much treatment as is needed to make the patient well and recognizing that overtreatment is not in the best interests of the patient."[9] The aim of therapeutic parsimony is the overall well-being of the patient, who should not undergo unnecessary procedures in pursuit of healing. Doctors must treat patients effectively without subjecting them to multiple interventions that may have unwanted side effects like toxicity, error, or discomfort. Simplicity is thus a matter of best care practices, in this case, as it protects patients from exposure to medical interventions that have diminishing returns or are harmful.

A third and final example of simplicity in biomedical ethics considers the impact of excessive antibiotic prescription on the human collective. Antibiotic stewardship recognizes that too frequent use of antibiotics builds up biological resistance. Eventually the human immune system may not be able to defend itself against simple bacteria and infection.[10] As antibiotics and antimicrobials are overprescribed and overused, more people stay ill and do not recover. Advocacy and awareness of antibiotic resistance and stewardship are taking place at the national level. For instance, the Centers for Disease Control and Prevention Stewardship program is "designed to ensure that hospitalized patients receive the right antibiotic, at the right dose, at the right time, and for the right duration."[11] Antibiotic resistance is becoming a major concern of health-care systems; stewardship and simplicity are authentic possibilities to address this concern in modern health care.[12]

In both environmental philosophy and biomedical procedures, simplicity is not merely a practice; it is also an attitude. Simplicity is a habitual mindset, an entrenched way of life, and a comprehensive philosophy that regards less as more, in many circumstances. Simple and sustainable medicine must reject the notion that bodies are always in need of the maximum medical interventions. Green Bioethics advocates for simplicity by reducing dependence on medical interventions through prevention and gradation. This attitude, or approach to life and biomedicine, must be thoroughly integrated into sustainable health care.

Reducing Dependence on Medical Interventions

There are seemingly innumerable aspects of health care that should reduce dependence on medical interventions. For example, doctors must reconsider when and how they prescribe pharmaceuticals, offer "routine" checkups, or suggest elective mastectomies and oophorectomies. Nearly any medical development, technique, or procedure the developed world has that the developing world does not, such as genetic therapy, nanotechnology, dialysis, Proactive skin care, and routine neonatal circumcision, could be evaluated for reduction. Two areas of health care that are often approached in the most resource-intensive manner are female infertility and obesity.

Female infertility and obesity are conditions that have significant personal responsibility attached to them. However, each may also indicate a clinical condition that may legitimately require medical attention. Both have numerous options for medical interventions. Because procreation and one's weight are deeply personal matters, ethical critiques—particularly from ecologists—are increasingly discouraged. In each of the representative cases of female infertility and obesity, however, a simplicity approach to addressing these conditions will conserve resources.

In the former situation, women can avoid dependence on medical interventions by taking active steps to prevent infertility, thus saving resources. Prevention of female infertility is particularly significant in the developed world, where fertility treatments are more frequently used.

In the latter example, when prevention of obesity is impossible, medical interventions should utilize low-tech, low-resource, economic solutions before moving to high-tech, complex, or resource-intensive interventions. The principle of simplicity challenges the assumption that health care has an obligation to intervene at the highest level for any preventable condition that could be considered a disease.

Prevention—The Case of Female Infertility

The United States is increasingly polarized over political issues, especially those revolving around government intervention and personal freedom. National incentives aimed at promotion of health through preventative measures face vigorous debate over "nanny state intervention" and personal autonomy. On the one hand, government involvement in preventative health—like mandatory seatbelts in cars, helmet laws, and the elimination of marketing cigarettes on television—is thought to save billions of dollars in devastating accidents as well as diseases.[13] More significant is that people who avoid preventable catastrophes have better lives. On the other hand, proponents of personal liberty, free from regulation, cite John Stuart Mill's dictum "that the only purpose for which power can be rightfully exercised over any member of a civilized community, against his will, is to prevent harm to others."[14] Thus, one's own health should be a private matter since it does not affect others. Particularly in topics that overlap with morality—such as sex, recreational drugs, and aid-in-dying—health-care debates balancing freedom and intervention are fierce. However, the environmental conservationist's position is rarely invoked when these issues are deliberated. Green Bioethics applies the principle of simplicity to female infertility by clearly assessing when and how it can be prevented. In many ways, the third principle of Green Bioethics is the most intuitive. When infertility is prevented, medical interventions are not sought, which will reduce dependence on health care.

Female infertility is often regarded as a personal, medical issue. Infertility can also be classified as a condition or state of being. The World Health Organization (WHO) defines epidemiological infertility as "women of reproductive age (15–49 years) at risk of becoming pregnant (not pregnant, sexually active, not using contraception and not lactating) who report trying unsuccessfully for a pregnancy for two years or more."[15] Self-perception of female infertility depends on the life plans of the woman. Women who do not want to be pregnant may welcome an infertility diagnosis; some women who seek biological children view infertility as an obstacle to be overcome. The health-care industry has declared infertility pathological and offers a plethora of medical interventions. Yet, infertility does not require a medical solution to what is a lifestyle preference—becoming a parent through biological reproduction—particularly because there is no somatic harm from not reproducing.

Medical interventions that are offered to produce pregnancies include low-resource mediations, like ensuring that intercourse occurs during ovulation, to more resource-intensive procedures like hormonal injections. Heavy use of medical interventions is seen across the gamut of assisted reproductive technologies (ARTs), including invasive techniques like intracytoplasmic sperm injection (ICSI), in vitro fertilization (IVF), three-person embryo creation, and gestational carriers. These medical interventions are instances of resource misuse since they do not cure or treat infertility. Prevention of female infertility draws on the principle of simplicity since it reduces dependence on medical interventions through bypassing the demand for fertility treatments.

In the developed world, infertility more frequently results in women seeking medical interventions for conception than in the developing world. Many factors like delayed procreation, obesity, and untreated sexually transmitted diseases lead to infertility.[16] Delayed procreation is often tied to extended artificial contraceptive use. When fertility is artificially suppressed for many decades, women may discover that they have "damage to reproductive organs from hormonal contraceptives and intra-uterine devices."[17] Infertility related to delayed procreation from extended synthetic contraceptive use could be prevented with non-hormonal contraception.

Additionally, Anne Dochin recognizes "in the global North, fertility is compromised by delayed childbearing" not necessarily linked to artificial contraceptive use.[18] Fertility declines with age; thus infertility related to maternal age can be prevented by voluntarily having children earlier. This has an additional benefit of preventing fetal complications associated with advanced maternal age, like greater exposure to environmental hazards.[19] Prevention cannot ensure that all women will be able to have biological children when they want to, but it will reduce the perceived need to utilize health-care resources for what is a normal and predictable result of certain chosen lifestyles.

Delayed procreation is largely a recent phenomenon due to the confluence of artificial contraception, economic and educational opportunities for women, and restructuring of traditional family models. Most people would view these as evidence of social progress. But now, many heterosexual, sexually active middle- and upper-class women in the industrialized world face competing goals such as completing one's education, career advancement, and procreation. Procreation is electively delayed in favor of other desires. Yet, it is a false dilemma to propose that women must either have children early and disrupt extensive education, or delay procreation and risk lifestyle infertility. These are not the only options.

Adoption circumvents pregnancy and sometimes the extra time that is needed to raise young children. The option to not have progeny, or have less than desired is another avenue for balancing reproduction and participation in the workforce. Women might also mitigate some of the tensions between career and procreation by finding a partner who will raise children as a co-parent. Many developed countries are still enormously sexist and expect that men will not participate equally in child-rearing, thus making women disproportionate caregivers by default. But, until men take an equal share of the burden of raising children, many women in the global North will continue to delay childbearing until they are established in their careers or finished with their education. Ideally, women would be able to make a truly free choice about if and when to have children naturally, along with how much education to undertake. If prevention of female infertility is to be a priority, then society must address structural sexism, which includes forcing the physical ramifications of artificial contraception and social result of child-rearing on women.[20] Instead, society has relegated approaches to infertility to health care, with dramatic implications for natural resource use.

Health-care workers and affiliates can also promote simplicity through prevention of female infertility. Health-care professionals have an obligation to make women aware that many forms of infertility are preventable. Public-health officials should be explicit about the lifestyles that contribute to infertility. As a social issue, health-care campaigns aimed at prevention of infertility would be useful, parallel to the public service announcements on life-threatening diseases like HIV, heart disease, and cancer. Since medical reproduction is within the realm of health care, it can be subjected to policy, which can also address prevention.[21] By preventing female infertility instead of depending on medical interventions, health care can become sustainable.

Prevention of infertility and natural conception still have the troubling outcome of population growth and carbon emissions expended on numerous resources from medical interventions during pregnancy and childbirth.[22] Therefore, any attempt at articulating a position that might appear to be pro-natalist—such as infertility prevention—must declare that the reason for encouraging natural conception is not because motherhood or procreation is beneficial for women, or even a morally acceptable decision in the presence of larger social concerns. "Procreation is no longer an unambiguous good, if it ever was."[23] Rather, prevention of female infertility would be advocated only to avoid medical interventions.

Infertility often leads to unnecessary health-care interventions. Treating preventable conditions contributes to the resource impact of health care. Thus, prevention is a first way of enacting simplicity and reducing dependence on health care. If conditions or diseases cannot be prevented, then a subsequent approach to simplicity may be applied within the third principle of Green Bioethics. Gradational provision of medical developments, techniques, and procedures may also be considered part of simplicity.

Gradation—Approaches to Obesity

Using Green Bioethics, approaches to managing and treating clinical diseases should rely first on prevention and then on gradation. Gradation is similar to therapeutic parsimony in its objective of using few resources, yet different in approach. While therapeutic parsimony often applies only to the lowest level of medical intervention, gradation fully integrates therapeutic parsimony on each level of medical intervention. Low-tech, low-resource, economical medical treatments with proven favorable outcomes should be pursued before escalating to complex, resource-intensive, or expensive procedures. Using gradation in medical developments, techniques, and procedures will reduce dependence on medical interventions. This is true whether the medical solution addresses a disease that has significant genetic factors like type 1 diabetes and some types of heart diseases and cancers, or a condition that is largely preventable, like female infertility or obesity.

Obesity costs the United States $300 billion per year.[24] For purposes of discussion, "obesity" will be an all-encompassing word for the range of conditions that indicate surplus body weight or body fat, including "overweight," "obese," "morbidly obese," and high body mass index (BMI)—a calculation of weight relative to height. Obesity, as such, is associated with a plurality of medical issues that may require medical treatments and, thus, is a case study in the application of simplicity.

While being "overweight" does not necessarily make one "unhealthy," there are a multitude of health complications that are associated with obesity.[25] Hypertension, stroke, cardiac disease, infertility, high cholesterol, and type 2 diabetes—which was formerly called "adult-onset" diabetes, but has been renamed because of the prevalence of diabetic and prediabetic children and adolescents—are included among obesity-associated comorbidities.[26] Obesity-related health conditions are not random; they are connected with well-known facts about the effects of excess body weight, body fat, and overconsumption of macronutrients.[27] Internationally, obesity is generally considered a disease of affluence that a majority of developed world citizens have chosen.

Inexpensive, high-calorie offerings and widespread accessibility of food from vending machines, grocery stores, restaurants, and supermarkets create the conditions for obesity to persist in many countries. Socioeconomic and demographic factors also partially determine obesity and health outcomes as well.

For instance, obesity-related conditions prevail in low-income areas in the United States, which are associated with race and educational attainment. Low-income areas are also characterized by less access to leisure and recreation.[28] Even so, health disparities, including those related to weight, are present in countries that mitigate the social determinants of health through socialized health care.[29] This indicates that obesity is generally a personal choice, within one's control, that is erroneously treated as a medical problem.

Obesity characterizes millions of people of all ages. The *Lancet* has thus recognized that obesity is not only an individual concern; it is also a public-health concern that burdens health-care systems.[30] Public-health measures have tried to address obesity on a national level by suggesting policies to improve nutritious food and the structure of urban environments, and to increase investment in population obesity monitoring. These initiatives save taxpayers money and medical resources, but have little public support in the United States.[31] It is contrary to the conventions of the modern health-care industry to implement systemic measures in order to reduce dependence on medical interventions. Resistance, especially in the United States, comes from the objection that preventative health measures interfere with liberty. Individuals demand "autonomy" at any cost, and reject health structures aimed at maintaining a healthy weight through balanced food consumption and adequate exercise. For instance, the initiatives seeking to decrease the consumption of supersized, corn-syrup-based, carbonated beverages are vociferously objected to.[32] Simultaneously, youth and childhood obesity are reaching epidemic proportions and inaction persists.

In the United States alone, 17 percent of youth are obese.[33] In England, almost 10 percent of four- to five-year-old children and 20 percent of ten- to eleven-year-old youths are obese.[34] Despite this growing concern, even policies that might prevent obesity in youth and children stall, since individuals prefer deregulation to sensible restrictions. The San Francisco's 2011 Healthy Meal Incentive Ordinance, which prohibited McDonald's from providing free toys in Happy Meals, caused a protracted debate. Adults decry the "surveillance" of children's diets, insisting that schools need not yield to nutritional standards.[35] Overreaction to intervention in youth obesity likens school programs that promote health and exercise to Michel Foucault's *biopouvoir*—a term that suggests national-level coercive force on citizens.[36] Americans have aggressively fought to maintain their—and their children's—consumptive, decadent eating habits, which they equate with freedom. Yet, resistance to public-health initiatives frustrates implementation of preventative measures.

While personal and public health measures are often rebuffed in favor of "autonomy," these intercessions in weight management result in better, healthier lives—measured in quantifiable outcomes like quality-adjusted life years (QALY) and infrequency of visits to physicians due to ill health. Rejecting preventative health-care measures demonstrates a lack of foresight, along with a fundamentally anti-conservationist perspective. This leads to the health-care industry taking responsibility for treating obesity-related conditions, instead of personal or social prevention, which is the simple solution.

In the rare case that obesity-related conditions cannot be prevented, then the simplicity protocol uses gradational approaches with lower-level, conservation-oriented solutions before high-tech, resource-heavy medical interventions. The health-care industry should not be the first stop on the road to losing weight. Thus, personal responsibility is the first level of gradation. Responsibility occurs after prevention has failed. This first level rarely uses medical interventions and has little resource impact. In the case of obesity-related conditions, responsibility appears in weight loss.

Weight loss is based on thermodynamics. When calories are consumed in excess of daily needs, weight is gained. When fewer calories are taken in than what is needed to sustain vital function and additional activity, weight is lost.[37] Those seeking to reverse the health effects of obesity can benefit from a reduction in calories.[38] The first-level approach of a modification of diet instead of reliance on pills, procedures, and medical interventions will support environmentally sustainable health care since it requires few, and oftentimes less, resources than are currently being used to maintain surplus weight.

Another first-level approach is exercise, which is free, available to all except the very infirm, and requires no extra resources. One does not even need to join a gym or fitness center to derive the benefits of walking, running, basketball, or soccer. The World Health Organization has suggested exercise for the sake of health and well-being as well.[39] Individuals who wish to avoid medical interventions in lifestyle obesity can take responsibility for their own weight through attention to caloric consumption and physical activity. The proactive, responsible, first-level approach to simplicity reduces dependence on medical interventions and conserves resources. Beyond diet and recreation, other first-level solutions to obesity-related health conditions abound.

In 2013, the cofounder of the Hastings Center, Daniel Callahan, suggested carbon-neutral strategies to address the obesity epidemic in the United States, like alterations in individual lifestyle habits, stigmatization, and social pressure.[40] These are environmentally sound solutions, but could be considered unethical in other moral spheres. Physician Garry Egger proposed carbon taxes as a way of reversing the expanding waistline of the world.[41] This would incentivize people to stay within the bounds of healthy weight or face financial penalties, but would disproportionally affect poor individuals.

After first-level solutions are thoroughly exhausted, medical intervention to treat obesity-related conditions must distinguish between medical developments, techniques, or procedures that address the health consequences of obesity and medical developments, techniques, or procedures that only address the aesthetic ramifications of corpulence. Plastic surgery, liposuction, and body contouring are in the latter category. Liposuction can spot-reduce fat, but unless the person modifies obesogenic environments and habits, they will regain the weight. Cosmetic interventions are offered under the guise of health care, but should not be considered medical solutions to obesity, since they do not treat clinical issues like blocked arteries, angina, or blood sugar levels. If medical interventions for obesity-related conditions are indeed required and clinically indicated, gradation is still warranted.

Gradation recognizes that a ratcheting of treatments may be necessary since first-level solutions may not suffice. For some, glandular conditions prompt obesity and only medical intervention will be effective. For others, the effects of obesity can no longer be managed with diet and exercise alone. There are multiple second- and third-level approaches to medical intervention in obesity-related conditions.

Second-level medical interventions for obesity are noninvasive. Resource differentiation and efficacy will vary. For instance, pharmaceuticals, which lower cholesterol levels and treat other obesity-related conditions, can be effective if the treatment plan is adhered to, but they have an enormous carbon footprint. Psychotherapy for binge eating or depression—both linked to obesity[42]—has very little resource use, yet success levels may be low. Furthermore, health insurance "incentivizes structures [that] typically reward surgical procedures over counseling,"[43] so it may not be an option for individuals.

Clearly, some medical interventions take more resources than others. Third-level medical interventions require technological specialization, with more resource use. These interventions also tend to be highly invasive. By way of example, tongue-mesh surgery utilizes a patch of metal sewn onto the tongue, making chewing difficult and painful, thus leading to a reduction in caloric intake.[44] The tongue mesh is removable, utilizes some resources, but is a temporary solution that must be continually replaced. Gastric bypass surgery is a one-time procedure that can drastically reduce weight, but has undesirable risks and sequellae such as possible lack of nutrition absorption, which could require further medical attention.[45] Another resource-intensive treatment for obesity is the Food and Drug Administration (FDA) approved Maestro Rechargeable System, which includes "a rechargeable electrical pulse generator, wire leads and electrodes implanted surgically into the abdomen"[46] to regulate appetite. This requires significant resources, and long-term effects have not been studied. A further highly invasive medical intervention to treat obesity-related conditions is an intragastric balloon that is surgically placed in the stomach and inflated to simulate feelings of satiety. The intragastric balloon is offered to adolescents as well as adults.[47] Each of the aforementioned treatments use varying degrees of resources. However, it is important to note that there are very few, if any, environmental externalities in preventing obesity or taking responsibility for weight. With added pressure on the medical system to address what could be avoided, the carbon impacts of obesity-related conditions proliferate. Utilizing first-level, simple approaches to treatment instead of depending on medical intervention are, in contrast, sustainable.

Obesity and related conditions are among the most pressing issues facing the developed world. Medical and natural resources are unnecessarily being put towards a condition that could be addressed on a number of nonmedical fronts—like prevention and gradation. Yet, reversing weight gain with sustainable solutions like calorie reduction and increased activity is not proffered, but rather energy-intensive pharmaceuticals, aesthetics, and surgery. When obesity-related conditions are treated, health care should use a gradational approach with first-, second-, and third-level medical interventions.

Both female infertility and obesity are preventable conditions. Bioethicists and ecologists should put efforts into awareness and prevention of these conditions, which are unnecessarily treated by health care. In the atypical case that female infertility or obesity cannot be prevented, a gradational approach to treatment can unfurl. This includes an attempt to reverse the condition instead of immediately treating it by medical intervention. Then, treatments that begin with first-level approaches should be utilized, only increasing in intensity, invasiveness, expense, or when absolutely necessary. While female infertility and obesity have been highlighted, many conditions have the potential to utilize the twofold approach of prevention and gradation. Medical and natural resources will be conserved if preventative and gradational approaches to medical interventions are employed.

Simplicity and Sustainable Health Care

Current practices that apply the principle of simplicity in health care may be difficult to observe because they are premised on absence: lack of treatment, lack of doctor visit, lack of prescription. Simplicity in action may also be affected by numerous variables that result in people abstaining from medical care, including personal responsibility, availability of health care, and financial limitation. It is far easier to quantify who uses health care, track number of doctor's appointments, and record pharmaceutical prescriptions. However, there are some visible examples of the intentional application of the third principle of Green Bioethics in action, effectively resulting in sustainable health care.

Simplicity is apparent in some cases of terminal illness and in certain clinical practices. In cases of terminal illness, the Natural Death Movement embraces simplicity by reducing dependence on medical interventions. In clinical practices, the Kimberton Clinic emphasizes simplicity through prevention, and then gradational medical treatments, termed "sustainable medicine." Both are premised on the inherently conservationist notion that while *care* should always be a component of health care, *treatment* may not be necessary.

Natural Death Movement and Reduced Dependence on Medical Interventions

Human beings have a 100 percent permanent mortality rate; all will succumb to death. There is no treatment or cure for death. Death is the natural terminus of every person's life, but some citizens of prosperity intentionally seek an interventional, resource-intensive death. The laborious use of health care in the last stage of life includes extensive employment of artificial nutrition and hydration, intubation and other forms of artificial life support, electrocardiography (ECG) monitoring, and finally, the inevitable expiration under medical supervision. Health care can and should address pain at the end of life through palliative care. However, numerous medical interventions during the dying process place a tremendous burden on health-care resources with minimal or no clinical benefit.

A humane and ecological approach to death that supports the dying person, and her family, encourages minimal invasion when the end nears. Instead of numerous medical interventions, the converse approach is to simply pass from life. Indeed, a nonmedical death has been the standard—and only—option for the vast majority of humans throughout the world. This environmentally sustainable approach can still be achieved in the modern world in a number of ways, like hospice and palliative care.[48] A further option highlighting simplicity is the Natural Death Movement, which reduces dependence on medical interventions while also providing the utmost comfort and care to patients.

In the United States, the Caring Advocates group, spearheaded by Stanley A. Terman, promotes natural death and the means to achieve simple and pain-free expiration. Carol Taylor and Robert Barnet paraphrase the Caring Advocates Natural Death Movement by remarking,

> Natural dying, like natural child birth, does not depend on high tech medicine, and it requires even less skilled assistance for nature to take its course. When

our brains can neither understand how to eat nor appreciate food, natural dying lets three things occur: 1. Cease manual assistance with oral feeding (as ultimately provided by skilled personnel), 2. Withhold/withdraw all life-sustaining treatment, and 3. Provide the best possible comfort care for a peaceful transition.[49]

Natural death thus removes dying from the medical industrial complex and focuses on a good death, acknowledging that death is inevitable.

Caring Advocates promotes natural death and encourages people to discuss end-of-life wishes before terminal illness or medical emergencies occur. The Caring Advocates organization offers "My Way Cards for Natural Dying" conversation-starting cards, which act as a step-by-step guide to living-will decisions.[50] In the Natural Death Movement, the highly interventionist default that one must do "anything necessary" to maintain vital functions is not assumed. Rather, following patient objectives and goals that include a timely and peaceful transition, complete relief from severe pain and suffering, avoiding prolonged medical harm, and alleviating the burdens of cognitive deterioration are promoted.[51]

Philosophies and practices akin to the Natural Death Movement are being advocated worldwide. Ian M. Seppelt, documenting end-of-life procedures from Australia, has discovered that "families who want 'everything done' . . . can usually be easily turned into wanting 'everything reasonable done with good comfort care and without invasive therapies or intensive care admission.'"[52] A medically laborious death is not the way many people envision their final end, but they inadvertently find themselves on a treadmill of hospital admissions, last-chance surgeries, and intensive-care-unit stays until death overtakes them. The Natural Death Movement provides a model of simplicity in health care that avoids dependence on medical interventions.

The Kimberton Clinic and Gradation

Richard Fried, MD, founded Kimberton Clinic in Phoenixville, Pennsylvania. Kimberton Clinic advocates economically and ecologically "sustainable medicine." Their philosophy of sustainable medicine is so striking that in 2014 the

Kimberton Clinic was profiled by the American Medical Association's *Journal of Ethics*. Associate editor Phil Perry observes, "At Kimberton, preventive medicine and sound prescribing are emphasized to eliminate overuse of antibiotics and psychopharmaceuticals."[53] Prevention is the clinic's first step in their model of sustainable health care. As a second step in sustainable medicine, Kimberton Clinic cautiously utilizes medical developments, techniques, and procedures when necessary.

Instead of falling into the consumptive cycle of "routine" use of scarce medical resources, the clinic approaches medical care using a gradational approach and offers individualized services that are sustainable.[54] They describe their philosophy by stating,

> While all up-to-date diagnostic services are utilized (CT, MRI, laboratory tests etc.) they are ordered with careful consideration, and not as a shortcut for thoughtful history-taking, physical examination and doctor-patient conversation. Thus, medical and therapeutic care is individualized, the opposite of a "cookbook" or "drug of choice" mentality. Our striving is to do the right thing at the right time for each patient.[55]

Kimberton Clinic demonstrates that simplicity can be achieved within a health-care establishment, with a balance of prevention, restrained prescription, and limited medical treatment. The model of Kimberton Clinic strikes the correct balance between use and overuse, and ultimately conserves resources. This approach to simple health care is an example for larger medical clinics to follow, especially those that tend towards immediate medical interventions.

The Kimberton Clinic serves patients of all ages and illustrates simplicity by reducing dependence on medical interventions. Kimberton demonstrates that patient-centered health care can thrive on sustainable medicine. Providing individualized treatments that respect the equilibrium of nature and resource use are more satisfying for the physician and patient as well. Philosopher Mary O'Brien observes that "it is a central contradiction in the lives of health care workers that there is a gap between what they think they ought to be doing—promoting well-being—and what they are actually doing, which is depending on ill-being for their livelihoods."[56] Doctors do not want people to be sick, they want people to thrive; but a medical practice based on unnecessary prescriptions, overuse of marginally beneficial treatments, and complex, multistep procedures is burdensome to the patient who has to endure medical interventions. It is also onerous on the planet, which must provide these offerings. Sustainable medicine resists the constant deferral to medical interventions. Simplicity—inclusive of prevention and gradation—is a path towards sustainable health care.

Conclusion

Doctors and health-care facilities are "raising expectations of what an individual or family needs to live well,"[57] while keeping the patient-consumer dependent on medical interventions. Television commercials endorse "solutions" to every physical and mental dissatisfaction humans have. Visual cues are reinforced by social messages that stir desires to be competitive in a free market, whether that competition is by being the smartest, most attractive, most fertile, or thinnest. Instead of medical interventions as a last resort, or even one among equal options, it is nearly mandatory for those who can afford it. This further normalizes the use of medical interventions, while "non-medical answers . . . are even more marginalized."[58] With a preoccupation with medical interventions, simple approaches to health care—which first rely on prevention, and then use gradation in medical treatments—are overlooked. The current model of medical interventions is contrary to sustainable health care. Society must consider a different approach.

The basic idea of simplicity in health care is one of economic theory, which takes "the most effective means to a given end."[59] Economy seeks to successfully accomplish a goal with the least amount of resources. The goal in Green Bioethics is just and sustainable health care. The resources are medical goods and services. Simplicity encompasses satisfaction with what one has and recognizes that when medical needs are present they should be addressed in the least resource-intensive way possible. Simplicity does not equate to a blanket rejection of all medical interventions. People can and should access health-care needs, in alignment with global distributive justice. However, simplicity in health care also avoids medical waste through the twofold approach of prevention and gradation. Addressing the aforementioned examples of female infertility and obesity-related conditions are selected illustrations; many more could be put forth.

Simplicity is vitally important for sustainable health care. Green Bioethics can strive towards the sun like Daedalus—utilizing the best of medicine when necessary—without succumbing to the fatal hubris of Icarus, attempting to unnecessarily intervene in every aspect of human existence. Overuse of health care through medical interventions is a major cause of resource use, but simplicity can lead to sustainable health care.

CHAPTER 6

Ethical Economics

The American health-care system is predicated on fees for service, expensive drugs, and costly health insurance. Each aspect of health care, from pharmaceuticals to surgical tools, operating rooms to patient beds, diagnostic tests to patient meals, has a profit margin goal that must be met. By the 1920s, hospitals in America were deeply entwined with a market approach to medicine. Profit was beginning to overcome patient care as a primary objective of health services. Guenter Risse records that since the turn of the twentieth century, hospitals needed to "create effective 'pull factors' to lure patients into particular institutions. There was no longer any pretense: health had simply become another commodity to be purchased."[1] Today, the choices health care makes about providing or creating new medical developments, techniques, and procedures are based significantly in finance.

In order to ensure continuous rollover of customers, slick marketing campaigns and medical enticements lure patients. Private detox centers, lavish psychiatric wards, and boutique gynecological clinics are eager to accept health insurance or private payments for their services.[2] These resource-intensive endeavors ignore the need for sustainable health care, rather embracing profit and serving an elite clientele. At the same time, millions of people cannot afford basic health care and the medical gap widens every day.

While all medical services are subject to cost efficiency, the demands for luxury medical goods, in particular, places enormous, unnecessary burdens on natural resources. The fourth principle of Green Bioethics—ethical economics—states that humanism should drive health care instead of financial profit. If the medical industry makes decisions about which medical developments, techniques, and procedures to offer based on ethical economics, health care can move towards sustainability. Ethical economics is not opposed to capitalism, but medicine must not lose its primary mission of health and healing.

The Health Care System as a Business

With increasing frequency, health care generates revenue in the same way as any other business: by kindling desires for products, selling them with an extensive profit margin, and advertising for specialty services. Every branch of health care is complicit. Even investigational research scientists "can no longer be counted upon to pursue solutions to major societal problems in health and human welfare . . . Priorities are dictated by commercial rather than social needs."[3] Set within a context of ethical environmental sustainability, health care should be directed at medical developments, techniques, and procedures that are not overly expensive, resource-draining, or rare. Yet, a persistent preference for exclusive-access, high-price, resource-intensive treatments call the commitment of health care into question.

Profit in the Health-Care System

The health-care industry in the United States accounted for 16 percent of the gross domestic product in 2009, while pharmaceuticals contributed 1.9 percent to the gross domestic product in 2015.[4] The massive revenue was not coming from meeting health-care needs simply, within a system of global distributive justice. Indeed, a panorama of the nonmedical offerings provided under the aegis of health care demonstrates a fissure between profitable luxury medical goods and the goals of medicine. "Destination hospitals" offer maternity suites, flat-screen televisions are provided in private rooms alongside individual menus, and cosmetic add-ons may be undertaken while having necessary medical procedures.[5] "Hotel hospitals" proliferate, enticing "guests" (i.e., patients) with stone fireplaces and waterfalls in lobbies, room service, and nail salons.[6] Procedures that were intended for medical needs are now used for elective enhancements. New Food and Drug (FDA)-approved pills and procedures abound, bringing resource-intensive luxury offerings to the affluent. The scale of the health-care industry continues to expand and collide with the natural limits of the planet.

As health care moves away from the goals of medicine—which tend not to be lucrative—and invests in for-profit models of business, the concomitant result is a plethora of nonessential services that are offered under the umbrella of health care. Nonessential medical services like non-reconstructive plastic surgery, aesthetic dentistry, elective joint replacement, lifestyle pharmaceuticals, and many dermatological procedures increase resource use of health care. Since they utilize doctors, prescriptions, pharmacies, and medical centers, they are rightly called "luxury" medical services.

Advertisements for these elective procedures reiterate the lack of clinical benefit, since required medical treatments must be utilized in the course of cure and need no promotion. Very often, luxury medical services are attached to marketing campaigns, thus preying on the consumerist tendencies of the middle and upper classes. C. Ben Mitchell et al. indicate that once basic health-care needs are met, "the confluence of an ego-oriented culture sustained by social approval, peer example, and clever advertising produce a cascade of demand" for unnecessary medical treatments that raise standard of living.[7] Ultimately, these luxury goods utilize the medical label and are legitimized as a part of the health-care industry, even though they are driven by commerce. Two examples highlight profit in health care: personal loans for fertility treatments and the Intercytex Ltd. medical company. Both require scientific invention, trained doctors, and marketing tactics.

In cultures marked by a procreative imperative, the absence of children in a relationship can signal a deficiency of initiative, because it is assumed that if one wants children, he or she can have them. "Establishing oneself as climbing up, rather than slipping down, the evolutionary track" becomes a pursuit accomplished by having biological children.[8] When children do not arrive naturally, the medical business offers infants on demand through fertility treatments for entrepreneurial, high-power couples and singles.

Fertility treatments, inclusive of hormonal injections, in vitro fertilization, artificial insemination, pre-implantation genetic diagnosis, and gene editing are examples of health care selling reproductive services to wealthy customers. Fertility treatments and assisted reproductive technologies can be expensive. In the United States and elsewhere, they are occasionally covered under health insurance—usually with limitations on egg retrieval cycles, cost, or on the number of conceptions.[9] In the absence of health-insurance coverage for fertility treatments, the high cost combined with low success rates translate to couples and singles without children to show for their investment. Since the treadmill of fertility treatments becomes more difficult to dismount the longer one tries to conceive, banks have strategically identified a new clientele. Banks are now financiers of loans for fertility services. Their strategy is shrewd, particularly because people who use the low-success fertility business translate to repeat customers.

In 2010, a financial center in New Zealand marketed personal loans aimed at couples pursuing reproductive technologies.[10] ASB Bank New Zealand created a television advertisement that begins with a woman indicating to her partner that her pregnancy test is negative. As the commercial tracks the failed attempts at conception, and a presumed diagnosis of infertility, the man sells his car, contacts ASB Bank New Zealand, and is approved for a loan to fund reproductive technologies. The ad omits scenes of signing a loan agreement, paying for the fertility treatments, or the assisted reproductive procedures themselves. Next, the commercial shows confirmation of pregnancy from the doctor and the purchase of a new minivan equipped with three infant seats. In the final scene of the advertisement, the infants are displayed, swaddled by parents and family. The ad ends with the phrase "ASB: Creating Futures" and appeals to the roller coaster of emotions in attempted pregnancy, relief at fertilization, and joy at new birth.

Obtaining personal loans for medical luxuries highlights the stratification of health care into medical need and medical profit. The ASB marketing campaign, along with other glossy advertisements for fertility clinics, indicates that banks clearly view fertility treatments as a business opportunity; their customers can take out personal loans to be used in exchange for services of their choice. While fertility treatments do not save lives and are therefore medically decadent, there are no correlative loans—or advertisements for loans—for basic health needs, like vaccines, or even high-cost but lifesaving organ transplants that might impact longevity.[11] Only well-established couples can afford to petition a bank for a personal loan and more than double the size of their family instantly. The conspicuous consumption reminds viewers that the medical business is not for everybody. Rather, "positional consumption"—defined as "goods that have the characteristic of allowing us to 'position' ourselves socially with respect to our fellows"[12]—displays the social ambitions of a young, white, infertile couple with good credit. At the same time, it should be remembered that while financial centers are willing to offer personal loans for fertility treatments, they cannot guarantee a return on investment. This is not troubling to banks and health care, however. Profit is made regardless of successful pregnancy.

Multiple corporations garner revenue when couples and singles use health care for technological reproduction. While all medical offerings are "fee for service," certain types of medical developments, techniques, and procedures expand market growth without concomitant health benefits. Luxury medical services are for the advantage of inventors, stockholders, and interested parties; yet these financial conflicts of interest degrade the integrity of the health-care system and deplete natural resources. Other aspects of the health-care business retain ties to scientific advancements without the assistance of third-party loans. Such is the case with Intercytex Ltd.

Intercytex Ltd. is a UK-based company that self-describes as "focused on developing our lead product ICX-RHY to treat a variety of skin related problems including Epidermolysis Bullosa and scar contractures."[13] In addition to these dermatological conditions, Intercytex also provides treatments for acne scarring and baldness.[14] The latter are not medical conditions, and the treatments offered are cosmetic in the clearest sense of the term. Intercytex promotes its products through the self-produced *Tressless News*, which publishes articles on "medical breakthroughs" for baldness. In 2006, the company made a bid to join the London Stock Exchange.[15] Customers can also follow Intercytex on Twitter.

Intercytex has been marketing its products for a number of years. Both scientific and medical investments have seeded the growth of this company, which treats both physical needs and superficial characteristics. This dual-purpose use of this health-care company reiterates, "Medicine is no longer just an art and a science. It is clearly a business now."[16] Luxury medical goods, such as the ones promoted by Intercytex, are sold to consumerist markets. Health care, in many parts of the globe, has succumbed to a capitalist mentality that requires expansion instead of equilibrium by creating artificial needs.[17]

The distinction between medical necessity and luxury medical good is obliterated when a prescription is tied to a clinically unnecessary drug, or when garnering a referral for a doctor specializing in "aesthetic" medicine, or when paying out of pocket for a cosmetic treatment that utilizes medical knowledge. Indeed, the medical industrial complex is, in part, so profitable because lavish items are available with very few regulations.[18] In addition to profit in the health-care system—that is, the physical structures tied to patient care—profit reigns in the pharmaceutical industry.

Profit in the Pharmaceutical Industry

The health-care system is a business with numerous lucrative offshoots. We are living in a pharmaceutical era, when nearly 70 percent of Americans take at least one prescription drug.[19] Pharmaceuticals accounted for 39 percent of the 546 million metric tons of carbon dioxide in the U.S. health-care sector in 2007.[20] Statista reports that in 2001, worldwide revenue from the pharmaceutical industry was near 390 billion dollars. In 2011, revenue had nearly tripled to over 963 billion dollars. By 2014, global pharmaceutical revenues reached one trillion dollars.[21] As generic medications splinter from name brands, the growing market for pharmaceutical profitability is established. Pharmaceutical companies have vested interests in maintaining patents, developing personalized drugs, and marketing pharmacogenomics. Without the development of new pills for new "diseases," financial profits will languish.

In 2005, Marcia Angell documented the shift from the development and production of medically necessary dugs to medically luxurious drugs. She revealed, "Profit motive has corrupted medical policy, practice, and research. Facing an actual downturn in innovation and in the development of new products, pharmaceutical companies are desperate to maintain their incredible profits. They achieve this by maintaining monopolies on drugs, introducing new, virtually identical drugs, and new slightly less efficient drugs, plus bribes, and advertising."[22] Angell's statement is confirmed by simply glancing at the 1,280-page *Approved Drug Products with Therapeutic Equivalence Evaluations*, also known as the "Orange Book," in reference to its jacket-cover color.

The "Orange Book" lists names, dosages, and ingredients in Food and Drug Administration–approved pharmaceuticals.[23] In many cases, the drug compounds are incredibly similar or identical, but produced by competitor companies. Many of the pharmaceuticals indicate a saturated market that represents stakeholders, inventors, chemists, and physicians all collaborating to promote their own versions of the drugs marketed to private practitioners and the public.

It used to be thought that "perfectly personalized medicine would, by definition, not be worthwhile to drug manufacturers; [because] there is little profit to be made in developing a drug for a market of one."[24] But, in 2015, the FDA approved 3-D pharmaceutical printing,[25] which could make pharmacogenomics not only reasonable but also incredibly accessible and, therefore, lucrative. These demands will be driven by the convenient delivery system— swallowing a pill—and the pharmaceutical companies will be able to expand their customer base. Personalized medicine, complete with bespoke pills that utilize biomarkers as the basis for creating custom-tailored drugs,[26] has become an option that will generate new demands on industrialized world citizens in years to come. Simultaneously, the pharmaceutical industry will perpetuate its current offerings.

In 2014, the *Boston Globe* profiled people who consumed expensive pharmaceuticals to treat symptoms associated with chronic diseases.[27] Tracleer, the brand name for bosentan, costs over $1,100 U.S. dollars a month. Tracleer eases the symptoms of pulmonary hypertension. It does not cure or treat disease and is considered a specialty drug. The *Boston Globe* reported that one woman pays for her pharmaceutical, in part, with a grant. The pharmaceutical industry has long been criticized for marketing unnecessary drugs to medical consumers, as well as overcharging for essential medications.[28] The United States is not the only country grappling with expensive drugs.

In 2015, the Competition and Markets Authority in the UK accused pharmaceutical companies Pfizer and Flynn Pharma of overcharging for phenytoin sodium capsules, which lessen epileptic episodes by inhibiting seizures.[29] The National Health Service confirmed this accusation by revealing that their cost of purchasing phenytoin sodium capsules rose twenty-five times, from 2 million to 50 million British pounds in 2013. Epanutin (brand name) pharmaceuticals are a form of preventative medicine, since unexpected seizures may lead to accidents, personal harm, and even death. The drastic increase in the cost of purchasing the anti-seizure drugs indicate that they can be sold for less, but that profit margins are driving costs.[30]

Profitability not only determines the cost of drugs, it also obstructs certain "unprofitable" drugs from being manufactured. From 2013 to 2014, there was widespread transmission of the Ebola virus in West Africa. Although a vaccine for Ebola was discovered in 2005 after a previous outbreak,[31] production was halted because pharmaceutical companies stated, "there's never been a big market for Ebola vaccines."[32] The Centers for Disease Control recorded that over seven thousand human lives—mostly from poor African countries—could have been saved if the medication had been readily available.[33] Instead, thousands of vulnerable people languished and died from the devastating effects of this high-fatality virus because the vaccine was not lucrative for drug manufacturing companies. Had health-care systems and the pharmaceutical industry been concerned about humanistic health care rather than profit, human lives could have been saved.

Occasionally, outrage over the price of lifesaving drugs receives media attention.[34] After a quick investigation, drug companies usually reduce prices.[35] In other cases, however, the pharmaceutical industry has intentionally chosen not to manufacture and distribute vital drugs to the poor in developing and developed countries.[36] The current environmental crisis, coupled with growing awareness of the need to secure global health care prompt ethicists to reassess the legitimacy of an economically driven pharmaceutical industry. Profit does not need to be the primary motivation in business. All sectors of commerce— the health-care industry included—have the ability to change their trajectories.

Ethical Economics

In ecology, there has been a growing corpus of literature on sustainable economics.[37] Green Bioethics draws on the concepts of sustainable economics to provide a pathway to ethical economics in health care. Specifically, ethical economics first identifies the appropriate role of profit in businesses as one, among many, values. Second, ethical economics inverts financial paradigms by describing an "ecological economic" that prioritizes ecological sustainability over production and consumption. Third, ethical economics offers an alternative to the gross domestic product, which is the current measure of national prosperity. The foundations of ethical economics thus simultaneously refute the monolith of profit and provide an economic model emerging from environmental concerns. This section cannot and will not function as a full-blown economic analysis of market factors, business, or commerce. Rather, it aims to demonstrate that there are many different ways to overcome profit as the singular compelling influence in businesses.

Appropriate Role of Profit in Businesses

There is nothing inherently wrong with making money from business. Businesses are the customary economic model in modern life, as opposed to an individual bartering system or self-sufficiency. Businesses in the modern world provide livelihood to individuals and afford them the means to purchase life's necessities. Businesses also drive the economy, providing stability to countries and the ability to participate in a global market. Yet, businesses can also be corruptive, demoralizing, exploitative,[38] and inconsonant with larger human values, such as dignity, health, and sustainability. As Kevin O'Rourke affirms, "Profit cannot be the basis of any profession but must be considered a secondary and highly variable feature."[39] Businesses must exist for more than simply revenue; they must account for human entitlement to a decent existence and the limits of the planet as well.

It may be difficult, at times, to assess where a particular business ranks profit or conservation. Some businesses, such as Zara clothing, send their wares in recycled boxes. However, they also rely on seasonal turnover of fashion trends. Tesla offers cars that run on electricity, but relies on fossil fuels for the production of its automobiles.[40] David A. Crocker, a senior research scholar at the Institute for Philosophy and Public Policy at the University of Maryland, has developed a model that evaluates the respective weights given to economic-related consumption and ecology.[41] His fourfold criterion highlights the interconnection between purchasing power, individuals, communities, and the current ecological problems. Crocker maintains,

> A given consumption practice (and thus, the business that offers that good) may be justifiable or defective in one or more of four ways. First, it may be good or bad for the environment. Second, consumption may help or harm other people. Third, our consumption practices may affirm or undermine values and institutions deemed essential to our community. Finally, a consumption choice or pattern may be beneficial or detrimental to a person's own well-being.[42]

Crocker's model is appealing because it is multidimensional. It neither ignores the reality of profit that may come from consumption, nor sacrifices the environment for the sake of wealth. A concise assessment of the American health-care industry, based on Crocker's model, finds it clearly falling short of the first criterion; the American health-care business is detrimental to the environment based on the metrics of carbon dioxide emissions and resource consumption. The second criterion can be seen from two different angles. On the one hand, the American health-care business may help other people as individuals receive treatments and cures that assist in their flourishing. On the other hand, the current consumptive practices of the American health-care business harm others through climate-change-related health hazards. In criterion three, the American health-care business at first glance seems to clearly affirm values of health, healing, and well-being. However, going beyond a superficial interrogation, the American health-care business also flouts values like justice, health-care access, sustainability, quality of life, and frugality. Of course, the aforementioned might not be values that enjoy wide consensus in the United States. The American health-care business generally satisfies the fourth criterion, with the caveat that iatrogenic disease and medical error cause significant harm in health care.

Despite the shortcoming of American health care, a business that places profit to the side of its corporate plan—instead of the center—is feasible. The appropriate role of profit in businesses must be subordinate to, or at least balanced with, environmental sustainability. Ethical economics can further be explained by the concept of "ecological economics."

Ecological Economics

Ecology and economics are typically faced with two competitive discourses. The first blames the current ecological predicament on the capitalist economy. In general, when products are sold, resources have been used. When production outpaces the ability of the earth to replenish itself, or the ingenuity of humans to create more effective ways to use fewer materials, sustainability is jeopardized. Conversely, there is the idea that ecology itself can be profitable and need not diminish revenue. "Green-collar workers" are drawn to jobs that promise wealth and environmental justice. Energy companies, automotive plants, and architects have all capitalized on the environmental revolution. Sustainable products and designs are highly appealing to the middle and upper classes. The supermarket chain Whole Foods is a well-recognized example of the (albeit shallow) alliance of capitalism and planetary awareness. Ecological economics combines the concern over resource exploitation and desire to earn a living ethically, within a sustainable economic system.[43]

Ecological economics recognizes the needs of individuals within a community. These needs include the ability to access not only consumer goods necessary for life, but also what is essential for a meaningful existence in a clean, healthy environment. Ecological economics attempts a synthesis of sustainability and profit instead of a unilateral focus on financial gain.

In order for ecological economics to be effective, a true investment in altering the behaviors of businesses is necessary. This takes the form of both realizing that all people are interconnected, and questioning unlimited economic growth in the first place. Markets—like all entities—are subject to fluctuations. Retreat must be a part of an economic system that maintains equilibrium. Many people in the developed world have been living in an age of abundance for decades. Now we must accept an alternative.

In the early 1970s, Van Rensselaer Potter drew on Kenneth Boulding's aphorism summarizing the requirement for economic reorientation: "Ecology's uneconomic / But with another kind of logic / Economy's unecologic."[44] That is, the economy is harmful to environmental harmony. In order to implement ecological economics more fully, ethicists must offer an alternative to equating net income with national success.

Alternative to the GDP: Gross National Happiness (GNP)

Since its formulation in the 1900s, the gross domestic product (GDP) has been viewed as the primary barometer of economic "health" of a country. Dan O'Neill defines gross domestic product as "an indicator of economic activity. It measures the total value of all final goods and services that are newly produced within the borders of a country over the course of a year."[45] The GDP indicates the dollar amount of economic growth and also assesses market performance of a country. It relies upon businesses to generate revenue as the primary form of economic expansion. Businesses do this by marketing products to consumers, who then purchase the items, spurring sales and profit. The GDP is a cyclical measurement in an infinite feedback loop. Spending creates profit; profit creates production; production creates spending. Despite the frequent appeal by politicians and economists to the gross domestic product as the exclusive indicator of national prosperity, it cannot be seen as the primary goal of national success for a number of reasons.

Financial generation is not the ultimate standard of society and can only increase life satisfaction up to a certain point.[46] Moreover, Martha Nussbaum rightly points out that gross domestic product can be asymmetrically distributed among citizens, so it is an unreliable determinant of individual human flourishing.[47] In addition to giving an anemic impression of the social and emotional fitness of citizens in a country, the GDP "does not look at all at other human goods that are not reliably correlated with the presence of resources: infant mortality, for example, or access to education, or the quality of racial and gender relations, or the presence or absence of political freedom."[48] Nussbaum clearly describes national success beyond economics. Yet, the gross domestic product relies solely on economics to assess a country's well-being and does not consider quality of life. In response, an alternative to the GDP can guide ethical economics.

Gross national happiness (GNH) has been proposed as a different assessment to GDP. Utilizing gross national happiness would inherently be a more sustainable economic system, since income—and thus resource consumption—would not be the driving factor. The concept of gross national happiness originated in Bhutan in 1972.[49] Gross national happiness is an adjustment to the phrase "gross domestic product." Nussbaum explains, "One step up in level of sophistication [from GDP] is an approach that measures the quality of life in terms of utility. This would be done, for example, by polling people about whether they are satisfied with their current health status or their current level of education."[50] Happiness is, of course, difficult to gauge, but it can direct society to the articulation of standards of life that are nearly unanimous.

Like the GDP, GNH primarily focuses on one multifaceted aspect of life illustrative of welfare. Whereas GDP emphasizes economics in general—and includes manufacturing, purchasing, selling, and capitalizing on services—GNH underscores happiness and includes health, sociability, security, family relationships, and community. Gross national happiness places emphasis on the needs of human beings and the quality of human life instead of determinants of market prosperity.

To be sure, the GDP cannot be replaced with GNH; they assess different values. Rather, gross national happiness is instructive for imagining requisite structures for nonmaterial satisfaction of people and national success. GNH recognizes that quality of life is a significant component of human welfare. Oftentimes, there is a connection between commerce and resource depletion. Yet, a fear of drop in revenue stymies sustainable business policy. All businesses are susceptible to this mindset. Thus, ethicists must offer an alternative. Ethical economics must be the foundation of commerce. In the medical industry, the principle of ethical economics can be applied through humanistic health care.

Humanistic Health Care

A sizable number of citizens are unable to access health care because it is cost-prohibitive. In the United States, even after the Affordable Care Act, financially insecure young adults were faced with paying fines if they could not secure health insurance. Many were too old to qualify for parent plans, did not have jobs where their employers offered health insurance, or had to pay an enormous amount from their paycheck for, effectively, a service they would not use. Approximately 41 million Americans were still uninsured as of 2015,[51] requiring them to make difficult choices about health care, insurance, and medical treatment based on economics instead of need. Under these structures, the poor majority are left behind while the rich minority pay their way into personalized insurance plans that generate revenue. When health insurance costs more than a mortgage, and people go into debt because they cannot pay medical bills for chronic or acute illness, the fissure between the goals of medicine and the health-care business widens. Of course, health care and health-care insurance goes beyond basic needs.

The American economic system—inclusive of health care—often places revenue above the welfare of all people. A proper understanding of humanistic health care ensures that each individual person's medical needs are met. Humanistic health care can be achieved by acknowledging—and acting on—the human right to health care. The right to health care can be further defined within the United Nations' International Human Development Index (HDI). Focusing on medical developments, techniques, and procedures that serve humanism instead of profit will ultimately reduce the amount of resources used.

Human Rights

Currently, rights language permeates a variety of settings, often oriented at groups like women; workers; lesbian, gay, bisexual, and transgender (LGBT) communities; or racial minorities. In ecological ethics, some philosophers argue that rights should be attributed to animals and plants as well.[52] Termed "biotic rights," flora and fauna are accorded life of their own, based on noninterference. These rights can only be violated "when necessary (the 'last resort') and only to the extent necessary (minimal harm)."[53] However, many people take an opposite approach to nonhuman rights, which does not recognize the intrinsic value of nature.

The Aristotelian view has a long intellectual tradition that binds many modern citizens to the idea; as Francisco De Vitoria proposed, "Irrational creatures cannot be victims of an injustice (*iniuria*), and therefore cannot have legal rights . . . it is lawful to kill them with impunity, even for sport; as Aristotle says (*Politics* 1256[b] 9–25)."[54] Environmental commentary on the rights of nonhumans is gaining growing acceptance on an international level, paved in part by the work that has been done on human rights. Significant, however, is the recognition that all rights are tied to each other. The environmental approach to human rights recognizes the interconnectedness of all people and living creatures.

Human rights express a concept that has been broadly recognizable since the United Nations signed the Universal Declaration of Human Rights in 1948.[55] In society, the rights of the person, obligations to others, and the good of the whole vacillate back and forth, ideally creating harmony where all people can thrive. Since human rights are situated within a larger community, these rights become intelligible only in the company of others. Rights, properly understood, are neither a handout, nor a claim that can be made without context. Meghan Clark summarizes the concept of rights by stating, "A right begins with a person who is bearer of the right (subject) and includes a particular substance (object) that is claimed against another individual or group who has a correlative duty to respect this right."[56] Human rights can only materialize when tied to larger networks.

For example, it is appropriate to say that if humanistic health care includes the right to basic medicine, then one has a responsibility to pay a fair amount to the provider. Conversely, if one has a right to choose medicine as a career, then that person has the responsibility to practice it in a way that allows all people to access his or her services. Both claims rest on a larger supposition that people who work should receive a living wage and that those unable to work must have their needs, including health care, met.

Human rights thus require responsibility from others when providing obligatory duties. This prevents individuals from making autonomous demands to entitlements that could jeopardize common resources. James Nash clarifies: "Using the rubric of justice, rights and responsibilities are not commensurate ... but they are correlative: strict responsibilities to other beings exist because the others have just claims."[57] It should be noted that the merging of obligations and responsibilities to rights answers the objection to human rights being narrowly individualistic. Human rights, therefore, demonstrate give-and-take and must benefit the person without neglecting society.

Since society is the context for all of human life—including rights claims— it must be the arena in which human rights take shape and are employed. Particularly with the institutionalization of medicine and the need for health-care insurance in order to access health care, some have argued that any and all forms of health care are rights. The specifics of the implementation of this right are debatable. Some people believe that the right to health care includes distribution of medical developments, techniques, and procedures at the uppermost limits. This would surely bolster the standard of care for the poor, making the right to health care more than just an idealistic suggestion. Yet, maximum resources expended on health care would disregard the limits of the planet. Furthermore, an enormous burden would be placed on health-care systems and government budgets.

Another approach to health care in a world society might suppose "the rich should have only the same health care as the poor."[58] This is a minimalist approach to medicine that would undoubtedly conserve resources, but would violate human rights to basic existence, since hundreds of thousands of people have no health care whatsoever. In 2008, a *Wall Street Journal* editorial entitled "Is There a 'Right' to Health Care?" claimed that there is a right to health care, but it "is less important than the right to food, shelter and clothing,"[59] thus positioning the claims of medicine within larger human needs.

Humanistic health care asserts that health care is a human right, while balancing the limits of the earth with the needs of all people. Since health care is for the benefit of all people, it must not jeopardize other common goods like the fecundity of the planet or the health of people living on that planet. Since it is a right, it should not be so expensive that people cannot access it. Due to its broad appeal and pluralistic appropriation, the language of human rights allows a large number of people to assent to, protect, and implement the manifestation of rights. It is foundational for humanistic health care. Claims to human rights are legitimized within a community, but they cannot be the only goal of society, since they are thin manifestations of human existence.

United Nations International Human Development Index

Developmental assessments, like the United Nations' International Human Development Index (HDI), expand on the concept of rights by recognizing that a full life is more than entitlements. Rights may include access to certain goods and services—such as health care—but it cannot measure other aspects essential to life. The Human Development Index thus assesses health, education, income, inequality, gender (i.e., sex), poverty, work, human security, trade and finances, mobility and communication, environmental sustainability, and demography. Each category functions as its own index and is determined by several subcategories.[60]

Like gross national happiness, the HDI emphasizes multiple features of human life constitutive of well-being. For instance, the "health" Human Development Index is comprised of life expectancy at birth, adult mortality rates for men and women, deaths due to malaria, deaths due to tuberculosis, human immunodeficiency virus (HIV) prevalence in adults, infant mortality rate, infants lacking diphtheria, pertussis, and tetanus (DPT) and measles immunization, public-health expenditure, moderate or severe stunting, and under-five mortality rate. Since the Human Development Index makes a direct appeal to health, these measures identify, in part, humanistic health care.

The United Nations specifies that life expectancy at birth is one of the essential, measurable elements of human development within the larger category of health. Linked with life expectancy are other characteristics of social health, such as adolescent birthrate, health index, homicide rate, maternal mortality, overweight children, and under-five mortality rate.[61] Humanistic health care would address each of these interrelated issues. For example, contraception can reduce adolescent birthrates. Vaccines and primary care can bolster the health index. Public-health measures—which need to be tailored to each country—can reduce the homicide rate. In the United States, for instance, a serious discussion about gun control and homicide is long overdue.[62] Maternal mortality can be addressed by better access to health-care professionals and medical care. National nutrition education and encouraging exercise can diminish the number of overweight children. Finally, enhanced postnatal and maternal care can decrease the under-five mortality rate. These health interventions are neither high-tech nor expensive. This makes their lack both inexcusable and also simple to remedy.

The UN recognizes that true development must include tangible aspects of existence. Humanistic health care must focus on these issues. Attention to the basic needs of all people, such as those identified in the Human Development Index, reorients the current consumptive medical industry towards sustainable health care.

Ethical Economics and Sustainable Health Care

In pursuit of profit, health care offers more medical developments, techniques, and procedures than are necessary for human welfare. The proliferation of medical goods and services has an enormous resource impact without necessarily resulting in clinical benefit. Particularly when pharmaceuticals are considered, the impetus for money in health care leads to resource exploitation. Prioritizing humanistic health care through ethical economics will lead to sustainability in health care.

Ethical economics can encompass many approaches, and under numerous permutations. Distribution of free antiretrovirals (ARV) for the treatment of HIV in Brazil is a noteworthy example of prioritization of humanistic health care instead of profit. It offers its citizens free HIV medication instead of forcing the vulnerable to contend with an illness and pay high drug costs. Additionally, Doctors Without Borders (DWB) meets health-care needs of people worldwide, without ability to pay as a prerequisite for service. As a humanitarian organization, DWB addresses the international gap in health-care access by bringing its services to people living in "medical deserts." Both are models of ethical economics.

Distribution of Free Antiretrovirals in Brazil and Humanistic Health Care

When human immunodeficiency virus/acquired immunodeficiency syndrome (HIV/AIDS) was discovered in 1981, treatment options were ineffective, toxic, or inaccessible.[63] In the last two decades, antiretrovirus therapy/antiretrovirals (ART/ARV) became the standard of care for those infected with HIV.[64] In the developed world, HIV/AIDS is no longer considered a terminal illness, but rather a chronic condition. However, other parts of the world are marked by a lack of antiretroviral distribution, particularly in poorer countries. In 2004, the United Nations estimated that six million people needed antiretroviral drugs to manage and treat HIV/AIDS, with the majority living in developing countries. In these low-income countries, "more than 9 out of 10 people are not getting the medications that, in richer countries, have turned a fatal disease into a chronic one."[65] Prior to 2003, Brazil was in a similar situation.

In the early 2000s, "the importation of three name-brand ARVs—nelfinavir, lopinavir, and efavirenz—were consuming 63% of [Brazil's] budget for acquiring ARVs."[66] The government was faced with an enormous health-care burden that was financially unsustainable. Both socially and economically, Brazil confronted a dilemma. The nation could have eliminated ARVs from available medical offerings in order to save money. However, "concern for human rights combined with the urgent need for access to treatment by people with HIV/AIDS bolstered wider efforts to lower the costs of ARVs."[67] The Brazilian government reduced the cost of providing the drugs to its citizens utilizing an economically sound, humanistic plan.

First, Brazil negotiated with international pharmaceutical companies to lower the cost of some ARVs. Pharmaceuticals are not sold at cost, and there is significant latitude to reduce the price of drugs, if companies are willing. Second, Brazil began manufacturing other ARVs domestically. Jane Galvão records that "of the 15 ARVs utilized in the country in 2002, 7 were produced in local laboratories, either public or private, and the remainder were purchased on the international market."[68] Combined, these two tactics facilitated distribution of ARVs for the treatment of HIV/AIDS at no cost to citizens. While both strategies are effective models for reducing the expenditure of health care, the domestic production of ARVs is of particular interest because money remained in the country, creating a positive loop of commerce. This strategy was so effective that by 2007, the Brazilian government decreed the compulsory license of AIDS medicine and began the national production of a generic antiretroviral.[69]

In Brazil, there were "at least two important arguments from an economic perspective for maintaining free access to AIDS medicines: the impact of ARVs in reducing deaths and the significant reduction in hospitalization and treatment costs associated with opportunistic infections."[70] In addition to pharmaceutical HIV treatment, Brazil also supplied free condoms to reduce HIV transmission, thus further buttressing their economically sound, humanistic health-care program.

Brazil began manufacturing condoms for its people in April 2008 when "the first factory in the world to produce condoms with natural rubber from a native rubber tree was inaugurated in Xapuri (AC) . . . In 2009, 34 million condoms were distributed for free by the company, built with federal resources."[71] The proliferation of barrier forms of contraception is both wise and effective, as it saves on health-care costs related to treatments for HIV-affected people.

The production of condoms also leads to greater national economic self-sufficiency since the country does not have to purchase them from other countries. The Brazilian government's strategy of producing their own HIV medications and asking for a lower price on medication that could not be produced allowed citizens to benefit physically and financially from necessary medical care. The manufacturing and distribution of free condoms reiterate the humanitarian priority of Brazil.

As HIV/AIDS was recognized as a global crisis, impacting people from all walks of life, individuals and organizations began to petition for broader access to medical treatments. The Joint United Nations Programme on HIV and AIDS (UNAIDS), under the leadership of Peter Piot; the work of ethicists Jon Fuller and James Keenan; and the commitment of physicians like Paul Farmer highlighted the need for HIV/AIDS care worldwide.[72]

By placing humanistic health care above profit, citizens received the medications they needed, the Brazilian government found an economical and ethical way to provide an essential medical service, and revenue stayed in the country. Brazil's approach secured HIV treatment for citizens and subordinated profit to humanistic health care. Of course, HIV/AIDS is just one disease. There are numerous health-care needs that are financially inaccessible in the developing world. Doctors Without Borders illustrates ethical economics beyond one nation's territory.

Doctors Without Borders and Ethical Economics

Known as Médecins Sans Frontières (MSF) outside of the United States, Doctors Without Borders (DWB) traces its origins to 1968, when French citizens Max Recamier and Pascal Greletty-Bosviel volunteered with the Red Cross in Nigeria. The pair remained with the Red Cross until MSF was officially created on December 22, 1971. Doctors Without Borders/Médecins Sans Frontières is a humanitarian-based medical organization that does not charge for its services. DWB/MSF summarizes its work as "help[ing] people worldwide where the need is greatest, delivering emergency medical aid to people affected by conflict, epidemics, disasters or exclusion from health care."[73] Its medical services are not for profit, but for the benefit of humanity. DWB focuses on emergency medicine, while also addressing other basic needs.

Currently, Doctors Without Borders/Médecins Sans Frontières provides free health care to the developing world, with a focus on access to medicines; treatment of various diseases like Chagas, cholera, Ebola, fistula, HIV/AIDS, kala azar, malaria, malnutrition, measles, meningitis, mental health disorders, sexual violence, sleeping sickness, tuberculosis, and promotion of women's health.[74] These are health threats that the developed world generally does not encounter, but drastically impact the well-being of people in the developing world.

Health conditions that are prevalent in low-income countries often coincide with governments that are unwilling or unable to address these problems. National political security is often tied to quality of health care for citizens. Thus, DWB/MSF "was created on the belief that all people have the right to medical care regardless of gender, race, religion, creed or political affiliation, and that the needs of these people outweigh respect for national boundaries."[75] Ideally, each country would care for the people within its borders. When this is not a reality, organizations such as DWB/MSF intervene for humanitarian—not financial—reasons. Doctors Without Borders/Médecins Sans Frontières invests significant effort to ensure that even the very poor have access to treatments.

When profits drive the creation, distribution, and marketing of medicine, enormous resources are expended. However, when humanistic health care is placed above profit, medicine will be allocated with enormous clinical benefit and minimal resource use. Free ARV distribution in Brazil and DWB illustrate the fourth principle of Green Bioethics by placing humanistic health care before profit.

Conclusion

In this era of consumerism, medical patients are offered developments, techniques, and procedures under the promise of happiness, fulfillment of life projects, or even higher salaries.[76] Luxury medical goods expand resource consumption and drive the economy. Compounded with this mentality is the illusion that health-care institutions only provide what is medically necessary and thus even luxury medical goods have some health benefit. Jessica Pierce and Andrew Jameton observe, "If wealthy industrialized societies as a whole are unsustainable, then so are the health care systems housed by these societies."[77] The environmental crisis remains unresolved, and yet if consumers are willing to pay, policymakers are hesitant to regulate the market for the sake of conservation. The world—and America, in particular—is convinced that prosperity depends on economic growth, regardless of environmental costs. Yet, financial gain cannot be the sole determinant of national success since revenue alone is a weak standard of human flourishing.

Purchases made in the realm of health care are purchases nonetheless. Money does not need to be absent from health care, but it cannot be the primary force, particularly given global ecological concerns. In 1971, Van Rensselaer Potter counseled that "technological decisions should not be made on the basis of profit alone, but should be examined in terms of survival."[78] Every aspect of modern life must endeavor to reduce anthropogenic climate change. This requires an upending of the current medical business models, which presuppose that profit is the ultimate goal of a business.[79] Ethical economics thus brings together the humanitarian aspects of health care with the imperative for environmental sustainability. As with the other three principles of Green Bioethics, implementation will require motivation and concrete action.

PART III

Green Bioethics
in Practice

CHAPTER 7

The Green Patient

n many cities there is a growing interest in environmentally sustainable choices from food to fashion. Individual consumers are purchasing electric vehicles, sponsoring community-supported agriculture, and recycling their household waste. On college campuses the 350.org initiative—spearheaded by Bill McKibben—has gained traction.[1] While Americans and other global citizens are increasingly electing for ecologically aware consumerism, there are few options for environmentally sustainable health care. A principled, environmentally sustainable approach to the use of medical developments, techniques, and procedures through Green Bioethics will attenuate resource consumption without compromising necessary health care. The application of these principles could be made in a variety of ways. Suggestions in this chapter are offered by way of illustration.

First, distributive justice entails using medical resources that are generally available to all people. Individuals may voluntarily choose to abstain from medical developments, techniques, and procedures that only a small percentage of the global population can access. Second, resource conservation will ensue when individuals choose health-care needs from a position of relationality.

Utilizing medical developments, techniques, and procedures from an inter-connected—rather than egocentric—perspective will temper resource use. Third, simplicity occurs when individuals make an effort to prevent disease and take personal responsibility for their health. Salubrious lifestyles within one's control will reduce dependence on medical interventions. Fourth, ethical economics may be employed through financial sharing of medical expenses. Those who wish to use health care can accept surcharges on medical goods that would be put towards humanistic health care. The principles of Green Bioethics can guide health-care consumers in making more environmentally informed decisions as "green patients."

Distributive Justice: Allocate Basic Medical Resources before Special-Interest Access

Individuals in the West often have control over how much they utilize health care. Due to the rise in patient-directed health care, people are also given numerous choices about medical developments, techniques, and procedures. Everyone should have access to basic medical resources—regardless of the environmental impact. Yet, many medical developments, techniques, and procedures are only available for a select few, indicating special-interest access. When medical options are clustered in wealthy areas while other people suffer without basic health care, resources are quickly drained. Green patients interested in applying the principle of distributive justice use medical developments, techniques, and procedures that are only accessible to everyone in the world.

Of course, the average person is not an expert in patterns of medical distribution. Medical developments, techniques, and procedures are determined by, inter alia, country, region, available technology, and ability to pay. Total parenteral nutrition (TPN) may be offered in Los Angeles hospitals, but not in rural Kansas. Even so, the number of special-interest medical developments, techniques, and procedures that are accessible in many parts of the United States—and other industrialized countries—is staggering and, in some cases, easily discernable.

Nearly ten million cosmetic procedures are performed each year in the United States.[2] Cosmetic procedures, also known as "aesthetic medicine" or "plastic surgery," alter the appearance of a person's face or body in order to conform to a concept of beauty or attractiveness. Cosmetic surgery is additionally marketed as a way to upward social mobility.[3] Some cosmetic surgery is reconstructive, such as skin grafts after severe burns or cranial reconstruction after neurosurgery. However, non-reconstructive cosmetic surgery is unnecessary from a clinical point of view and flouts distributive justice since it is not widely accessible.

Common types of aesthetic medicine include cosmetic dentistry, blepharoplasty (eyelid surgery), face-lifts, breast augmentation, rhinoplasty (nose surgery), otoplasty (ear surgery), liposuction, and breast lifts. These procedures—which are unavailable to the majority of people in the world—are offered to placate the demands of medical consumers who have their general health-care needs met. Oftentimes, those who seek cosmetic surgery have multiple procedures done over many years. These medical developments, techniques, and procedures frequently dovetail with conspicuous consumption.

In 2002 alone, United States citizens spent $7 billion dollars on non-reconstructive plastic surgery[4] and had 9.3 million procedures performed in 2005.[5] Although most consumers in the developed world are not consciously attempting to medically outpace those in the developing world, there is a sense of competition for more medical procedures among those within industrialized countries. Particularly with an overemphasis on outward appearance and youthful appearance, cosmetic procedures demand more resources to sustain the artificiality that they create.

At the same time, medical clients are caught in the matrix of health-care policies that give the illusion that medical developments, techniques, and procedures are basic medical resources since they are covered under health insurance. David Crippen observes that oftentimes "consumers of health care are not the purchasers [of health insurance] and so have little motivation to assess cost versus value. More is always better, especially when it is free."[6] It is difficult to convince Americans otherwise, unless there is a strong ethical appeal to, for instance, distributive justice.

Jessica Pierce and Andrew Jameton thus implore "the world's wealthy consumer classes, who spend roughly 90% of all of the dollars spent on health care in the world, [to] be sensitive to ethical principles, suggesting that they should reduce their consumption of health care materials and services."[7] Since the ability of the planet to provide for its inhabitants is currently jeopardized, individuals should focus on using medical developments, techniques, and procedures that fit within a pattern of equitable health-care allocation. Distributive justice in health care is a necessary step toward sustainable health care.

Resource Conservation: Provide Health-Care Needs before Health-Care Wants

There are a plethora of health-care amenities from which to select, and it may be difficult to determine what type of medical developments, techniques, and procedures are environmentally sustainable. A description of health-care needs—namely, those that conform to the goals of medicine and support function and quality of life—was provided in chapter 4. Yet, health-care need can be difficult for individuals to ascertain for reasons such as the disputable line between function and enhancement and quality of life and standard of living. Thus, prior to applying the principle of resource conservation, a green patient can place herself in a mindset of relationality, which can reorient the medical consumer and ease the intellectually demanding job of parsing health-care need and want.

Relationality is integral to personal identity in many non-Western articulations of bioethics and recognizes the interdependent networks of life that are indispensable for all humans to have their needs met.[8] These networks, also known as communities, emphasize the impact that we have on others and the impact others have on us. Indeed, "environmental philosophers . . . appreciate individuals as strongly connected with all humans, creatures and the natural world in a cyclical flow of materials and energy."[9] Since humans are connected and dependent on others in a community, individuals must make medical choices acknowledging that the self is simultaneously part of a larger whole. That is, a patient should choose the developments, techniques, and procedures that meet her health-care needs within the community.

By way of example, the Amish people often deliberate within their community if they can afford to treat a terminally ill patient with aggressive but futile measures. They do this with the highest regard for the dignity and value of the person, balanced with the corporate needs of the group. Sometimes the community rejects extraordinary measures for the individual. While this should not be imposed on someone, the relational mindset makes rejection of medical wants—which include aggressive but futile and extraordinary medical treatments—an accepted part of life.[10]

Individuals should certainly protect self-interest in maintaining health, yet, if more people approached health care from a position of relationality, eventually the tide would turn and health-care choices would be informed by environmental values. Through this new sense of personal identity, there will be a reduction in the demands for health-care wants since the individual will consider her impact on the global community.

Health care in the developed world causes extreme resource use. The National Health Service notes that the health-care industry cannot continue to expand and that "current levels of growth [will need] to not only be curbed, *but the trend to be reversed.*"[11] Currently, the Western health-care system regards the individual as a monad who is unrelated to other people, other creatures, and the environment. This self-centered attitude encourages indiscriminate consumption of goods that raise one's standard of living. Utilizing medical developments, techniques, and procedures that are wants are seen as unproblematic because autonomous patients are told they can—and should—make health-care decisions in accordance with their own preferences, irrespective of others. Medical consumers in the developed world tend to evaluate only their own personal desires, not limitations to the ecosystem. As a counterbalance to this mindset, relationality emphasizes an inherently dignified individual situated in a community.[12] Internalizing relationality prior to assessing health-care need or want will lead to resource conservation.

Simplicity: Reduce Dependence on Medical Interventions

The industrialized medical industry is inundated with people who could have—through preventative measures—avoided medical interventions. Medical developments, techniques, and procedures that address the ramifications of detrimental lifestyle choices increase pressure on health-care systems. Medical patrons who wish to apply the third principle of Green Bioethics can reduce their dependence on medical interventions through prevention and responsibility. Prevention bypasses the need for medical intervention and, in turn, avoids the environmental pollution of drugs, treatments, and procedures that health care expends on preventable illnesses. Responsible individuals proactively take control of health and lifestyle choices. Here, the "green patient" is not a patient at all, but simply a person who is living a life that does not require medical dependence.

The ethical imperative to prevent disease and take responsibility for one's health has a long intellectual tradition. In *Nicomachean Ethics*, Aristotle describes the case of "a man who becomes ill voluntarily through living a dissolute life and disobeying doctor's orders. In the beginning before he let his health slip away, he could have avoided becoming ill."[13] Prevention of diseases within a simplicity model will lead to a more sustainable health-care system, since the absent individual translates to a reduction in medical resources; where there is no patient, there are no resources expended.

Of course, prevention is not always possible. Some diseases, such as cystic fibrosis, are hereditary and could not have been avoided by the person, although the parents could have taken measures to prevent transmission of the disease. Other conditions—including many forms of asthma—develop as a result of environmental pollution. Moreover, race, educational attainment, and access to health-care systems are also key determinants of diseases and disorders in the United States. Obesity, type 2 diabetes, and high cholesterol appear in African Americans in higher numbers than European Americans in the United States. These are avoidable conditions, but numerous interlocking factors such as systemic racism, poverty, and stressful life circumstances make prevention challenging.

In other cases, diseases are tied to broadly accepted lifestyles. Alcohol is a part of the socialization in many cultures and can cause cancer, liver failure, jaundice, and lead to alcoholism. People who drink excessively may lack coping mechanisms necessary to reduce stress, or are unable to access other forms of entertainment or relaxation. Alcohol consumption may also be tacitly endorsed by numerous pubs and bars in towns and cities. Yet, external factors do not absolve each individual from being responsible for his or her own health, when possible. Thus, prevention as a principle of Green Bioethics does not reprimand those that cannot control their circumstances, but rather encourages those who can.

Personal responsibility can take many forms that range from simple to complex. A part of responsibility in health care requires that short-term "sacrifices" in lifestyle—like eating well, exercising often, preventative skin care, stress management, and relaxation—take priority. More challenging changes might include ceasing to use tobacco products and other drugs, reducing alcohol consumption, preventing numerous pregnancies, and driving safely. Responsibility requires critical thinking about lifestyle choices and medical interventions as well.

Notably, nearly 70 percent of Americans use at least one pharmaceutical.[14] Many of these prescriptions are unnecessary and have non-chemical treatments. Instead of relying on medical interventions, changes in diet to lower cholesterol, relaxation instead of blood pressure medication, and exercise to combat depression are nonmedical, low-resource options. Reports on both the National Health Service in the *British Medical Journal* and the United States health-care sector in the *Journal of the American Medical Association* named pharmaceuticals as greatly contributing to the carbon footprint and resource use of health care. This is due to the fact that "pharmaceutical products with complex manufacturing processes [have] environmentally significant precursors . . . as well as complex and hazardous solid, air and water emissions, including toxic, infectious and radioactive wastes."[15] Medical consumers can frequently control the amount of medical intervention in their lives.

Autonomy—understood as patient choice in medical interventions—is often touted as the supreme principle of medical decision making. However, when individuals proactively take control of their health, they are most fully demonstrating their autonomy. Individuals retain more autonomy when they circumvent health care and do not depend on medical developments, techniques, or procedures. Although autonomy is highly prized in health care, this Western mindset ensnares individuals by allowing them to believe that they can do whatever they want to their bodies without repercussions.

Thomas Szasz rightly observes that Americans want "an authority that will protect them from having to assume responsibility not only for their own health care, but also for their behaviors that make them ill . . . politicians assure people that they have a 'right to health' and that their maladies are 'no-fault diseases.'"[16] Chimerical promises for chemical cures or lifesaving surgeries remove personal agency. This reactive approach to disease is an uphill battle that not only threatens massive resource use, it also damages the individual who may never find a cure for the condition that responsibility could have prevented. For this reason, Michael Yeo advocated for an "ethics of empowerment approach . . . that demands that people be assigned a 'prospective' responsibility for their own health, as opposed to a 'retrospective' one" as early as 1993.[17] Beyond the immediate health benefit to the individual, the prevention of illnesses and diseases is a significant avenue towards environmental sustainability in health care since it will reduce dependence on medical interventions.

When individuals are not responsible for their health, they use medical resources unnecessarily and the resources are depleted. However, prevention of diseases and conditions supports sustainability. These lifestyle changes will yield tremendous environmental results as people reduce their use of medical developments, techniques, and procedures. Prevention and responsibility radically undercut the current medical model of unsustainable Western medicine by opting out of medical intervention whenever possible.

Ethical Economics: Humanistic Health Care Instead of Financial Profit

Many aspects of life demand that we make positive ethical decisions, or at least refrain from inflicting further harm. Participation in structures that damage the environment—such as the health-care industry—can either be renounced altogether or changed from within. The latter approach to sustainable health care proposes sharing the financial burdens of humanistic health care worldwide.

In these cases, medical developments, techniques, or procedures that are clinically indicated would cost slightly more for the wealthy. The surplus income would then be used to fund general health care for those without. Individual medical consumers who wish to participate in ethical economics can share the financial costs of humanistic health care.

Ethical economics, as a principle of Green Bioethics, can be implemented by either centralized or decentralized financial sharing structures. In the former, the government directs excess funds from its citizens into humanistic health care. In the latter, the individual herself chooses the organizations to which she will contribute. Both methods mimic "free market environmentalism" (FME), where purchases are used to support sustainability,[18] or in this case, a sustainable health-care system. The United States Patient Protection and Affordable Care Act (ACA) is an example of centralized financial sharing that supports humanistic health care.

When the Affordable Care Act was passed in 2010, it was hailed as a long overdue step toward extending health-care insurance to millions of uninsured people. Higher monthly insurance premiums for the middle and upper classes gave health-insurance coverage to people in the lower class. The ACA also provided health-care insurance to those with preexisting conditions and other vulnerable populations. Although health-care insurance was more expensive for some, others found health-care insurance affordable for the first time.

To be sure, the structure of the Affordable Care Act was imperfect. A sizable number of citizens are still unable to access medical developments, techniques, and procedures because of their cost. The insured face high co-pays and deductibles that make them hesitant to access medically necessary procedures. Others with health insurance avoid medical care because of the fear of receiving a staggering bill. But many of the objections to the ACA would more rightly be directed at the artificially inflated costs of health care. The Affordable Care Act's primary objective was to compel the rich to pay more for their health insurance in order to redistribute health-care costs. Participation in a centralized economic system that protects the human right to basic health care—like the ACA—is one form of ethical economics.

In addition to governmental oversight of individual financial contributions to health care, decentralized financial sharing encourages individuals to contribute to the health-care program of their choice. Decentralized financial sharing of sustainable medical developments, techniques, and procedures gives individuals the opportunity to practice ethical economics on their own terms. This enhances individual accountability for community and environment, while also allowing preferences to be expressed in the medical marketplace. Planned Parenthood is one illustration of decentralized humanistic health care that depends on individual financial contributions.

Planned Parenthood is a national health-care organization that specializes in medical developments, techniques, and procedures and preventative services for low-income men and women. Under United States law, Title X allows Planned Parenthood "to supplement birth control, gynecological care, and other reproductive health services for women who cannot pay full price for health care services."[19] Planned Parenthood provides many services, including prenatal care, mammograms, diabetes and breast cancer screening, pap smears, and flu vaccines.[20] These comprehensive health-care needs have a free or sliding scale attached to them, indicating that an inability to pay should never prevent health-care needs from being accessed. Planned Parenthood's various health-care services are clearly humanistic.

When medical customers donate money to medical organizations that work from a constrained budget, they can be confident that they are giving to patient-driven health care, not profit-driven medicine. There are numerous examples of health-care organizations that are not-for-profit but still deliver humanistic health care and medical services to the poor. These include the United Nations Millennium Development Goals, United Nations AIDS, World Health Partners, and the World Health Organization. Organizations with constrained budgets must necessarily prioritize medical developments, techniques, and procedures that will have the most significant clinical impact on those in need of medical attention. Individuals can choose to contribute to these, and many other, associations that exemplify humanistic health care. Distributing the cost of medical care—whether through centralized or decentralized means—places humanistic health care above financial profit.

Conclusion

In the United Kingdom, the National Health Service instituted structures for reducing medical waste and carbon emissions; in the United States there are no such measures. In the absence of systematized initiatives, environmentally oriented action must occur on every level possible. This includes consumer choice. Through subsidiarity, which "asks not for the most local, but the most appropriate level of organization and response,"[21] people in all levels of society can commit to sustainable health care.

In many ways, environmentally sustainable health care begins with the individual. It is, ultimately, people who choose which medical developments, techniques, and procedures they will pursue. Health care exists for the person, and as long as people view health care as an unlimited commodity, it will be unsustainable. As a counterbalance, individuals can utilize the principles of Green Bioethics to guide their medical decision making.

CHAPTER 8

The Green Doctor

octors hold a unique position of privilege as respected, educated, authoritative healers. Their medical training and profession "opens doors for them to meet with elected representatives, influence policy through such means as letters to the editor, and act as spokespeople for important causes and media contacts for questions relevant to public health."[1] Physicians demonstrate dedication to human health and ethics in a number of ways. Doctors have ample opportunities to participate in Green Bioethics and significant latitude to integrate sustainable health care into their practices.

First, distributive justice advocates doctor redistribution through incentives, loan forgiveness, and policies that support doctors pursuing work in medical deserts. Second, doctors who value resource conservation can halt medically non-beneficial treatment. This would require institutional support to protect doctors from emotional and legal ramifications of physician-family disagreements. Third, simplicity can occur when gradational approaches to technological interventions are utilized. Simplicity is especially apropos when low-resource alternatives to treatments abound. Fourth, ethical economics encourages doctors to provide humanitarian medicine instead of financially lucrative nonmedical services. Physicians should use their medical training in a responsible way that only utilizes resources when they are necessary.

Distributive Justice: Allocate Basic Medical Resources before Special-Interest Access

Accessing general health care is virtually impossible for some people in the world because doctors are clustered in some areas rather than others. The World Health Organization examined doctor density globally and concluded that people living in developing countries often face a primary care physician (PCP) shortage. In 2014, many African countries reported less than 0.3 doctors per 10,000, or 3 doctors per 100,000 people. In contrast, the United States reported 25 doctors per 10,000 people, or 250 doctors per 100,000.[2] Fewer doctors result in worse health outcomes for potential patients, diminished lifespans, and risk of living with an acquired disability.

Women are disproportionately affected by this shortage due to the physiologically distinct features of female existence, i.e., pregnancy and childbirth. Placement of doctors is a precondition to accessing medical developments, techniques, and procedures. Thus, doctors can participate in the first principle of Green Bioethics by intentionally seeking opportunities to provide medical developments, techniques, and procedures in understaffed areas. Moreover, financial incentives—either nationally or internationally—can support doctor redistribution, which necessarily entails medical resource redistribution.

In some countries, doctors choose to go into medical school knowing they will be placed in underserved locations. In 1968, the Thai government imposed "compulsory contracts with medical students so that they had to perform three years of public work after graduation or face high fines."[3] This was a reaction to "external brain drain" that siphoned native doctors to the United States and the UK. Indeed, the *American Journal of Public Health* indicates that "staffing shortages, lack of specialist training in poorer countries, and the financial lure of the West have resulted in the migration of physicians and nurses from the mostly developing source countries to the more developed host or destination countries."[4] To counteract this relocation, predetermining the location of employment partially equalizes doctor distribution.

Furthermore, incentives provide doctors with a real choice about where they will practice instead of fewer options because of financial pressure to work in high-income, low-need areas. Future doctors can pursue financial scholarships from medical schools that require a commitment to work in medically underserved areas later.[5] Establishing a practice in medically underserved areas can redistribute doctors from global North to South, or within developing countries.

Investing money in fortifying the medical training of doctors in the developing world is another option. This would create jobs and skills using countries' own resources. A further option is accepting international medical students to be trained at developed countries' own universities and then encouraging them to return to their own country to work. Incentivizing physicians to work in underserved areas and fully utilizing available internal medical resources can result in doctor redistribution.

Adequate financial compensation is a significant component of doctor redistribution. Doctors should be compensated for their time and skill that are, oftentimes, of benefit to humanity. In countries where the salaries of doctors are not competitive, there is migration towards nations with lucrative remuneration.[6] Offsetting the incomes of physicians in the developing world via funds from the developed world would show solidarity with the poor and make medical practice in these areas more attractive. This could be done through alternate financial arrangements, whereby medical degrees in the developed world would have higher tuition, with the surplus money providing scholarships or subsidizing medical degrees in the developing world. Health care must address "the international context, asking whether and how justice can be well-served between developed and underdeveloped countries."[7] Developed world citizens are obligated to share a greater part of the human-resource burden of securing global health-care justice for all because they have already received significant medical, economic, social, and political benefits. Countries that produce high numbers of well-qualified health-care professionals must collaborate with countries that do not have the same advantages.

Remediating the unjust concentration of doctors worldwide requires that numerous structures support doctors who wish to use their training in places that suffer from medical scarcity. Doctors must also lobby for these changes and actively pursue paths that allow them to provide medical developments, techniques, and procedures according to distributive justice.[8]

Resource Conservation: Provide Health-Care Needs before Health-Care Wants

Green doctors can utilize the second principle of Green Bioethics—resource conservation—by providing health-care needs that align with the goals of medicine, provide function, and offer quality of life. Since doctors are often responsible for the amount and kind of medical developments, techniques, and procedures given to patients, they can focus on health-care needs instead of health-care wants.

As a first step, doctors could curtail providing elective surgeries that offer little to no medical benefit. In addition to exceeding the goals of medicine, elective surgeries are resource-intensive and often create additional medical problems. In one study, patients undergoing major elective surgery were at risk of developing postoperative complications because of either the surgery or the presence of coexistent medical conditions. Moreover, "between 65–74% [of patients undergoing a major elective surgery] had one or more medical conditions like heart disease, hypertension, pulmonary embolus, or diabetes mellitus."[9] These individuals, who were also in the later stages of life (between the ages of sixty-four and seventy-seven), were already heavy medical consumers and presented extraordinary burdens on the health-care system.

Elective surgeries and other medical developments, techniques, and procedures not related to health-care needs can be denied for several clinical reasons, such as the high risk they pose to the patient, the potential for death after surgery, and the priority of treating their other chronic conditions. The last point is particularly salient because elective surgeries are habitually offered in response to negligence of personal health and do not target the underlying condition, but rather act as a "workaround." Limited medical resources must be directed at the source, rather than the expression, of the ailment.

As a second step, doctors must be given more power to determine when a particular treatment is no longer beneficial and not be obligated to provide these treatments. Many times health-care providers in intensive care units provide medically inappropriate treatments at the request of the patient, or the patient's surrogate decision maker. Futile treatments in medicine can be defined as "unnecessary therapies, overtreatment, or therapies that will not achieve a desired goal."[10] Ethicists have pointed out that by avoiding futile treatments, resources could be saved, which could benefit other patients.[11] And yet, in a patient-driven medical marketplace, there is pressure on doctors to maintain futile treatments, especially when they are demanded by patients in the form of an advance directive, or by family members.

Since doctors are responsible for writing patient-care orders, they must take a proactive role in establishing when a treatment is no longer effective. Notably, in South Africa, "law does not place the burden of decision on the family, but rather on the clinician, as the decision whether a particular therapy is futile is deemed a medical decision, which the family cannot refuse."[12] Doctors have the medical skills and expertise to assess the potential for recovery and return to baseline function. Families and patients cannot compel doctors to provide treatment when it is not medically indicated; that would be contrary to best medical practices. Indeed, medically non-beneficial treatment in the setting of complicated grief can be "worse than futile."[13] Doctors must be given more latitude to practice in accordance with medical standards instead of bowing to unreasonable patient demands. In the case of elective, risky surgery and futile treatment, providing health-care needs before health-care wants aligns with best medical practices and leads to resource conservation.

Simplicity: Reduce Dependence on Medical Interventions

Green doctors may apply the third principle of Green Bioethics in a variety of ways. Simplicity encourages patients towards responsible habits, and advocating for preventative health-care policies can shape physician practice. When medical resources must be expended on medical developments, techniques, and procedures, simplicity reduces dependence on medical interventions through gradation. In the latter case, some people who present to health care for interventions have a multitude of first-, second-, and third-level options available to them. This is particularly true when the patient requests a cosmetic, clinically controversial, or elective procedure. Doctors can direct patients seeking particular medical services away from high-technological medical developments, techniques, and procedures and towards lower-level interventions.

For example, people who seek surgical intervention to enhance or confirm their gender identity often substantially use health care. This is true for cisgender individuals (a person whose natal sex aligns with their gender identity)—for instance a woman who seeks cosmetic surgery to exaggerate her breasts or narrow her waist—and also transgendered persons (a person whose natal sex does not align with their gender identity) who request sex reassignment surgery. The latter is the focus of this section, but much can be applied to the former case as well.

Sex reassignment surgery (SrS) is a term that describes the range of technological and surgical medical developments, techniques, and procedures by which a person's physical appearance and secondary sexual characteristics are altered to resemble that of the other sex. In the United States medical industry, SrS is one generally accepted treatment for gender identity disorder (GID)/gender dysphoria in transsexual and transgender people, who have been categorized as mentally disordered by the *Diagnostic and Statistical Manual of Mental Disorders (DSM)*.[14] Note, gender identity disorder (GID) was the term used in the *DSM*-IV, but it was removed from the later *DSM*-V and replaced with the term "gender dysphoria." The criteria for diagnosis are similar for both conditions, and it must also be remembered that many people feel some dysphoria about their physical appearance not related to gender.

A gender dysphoria diagnosis is often associated with a transgender identity. That is, transgender people have gender dysphoria, according to the *DSM*-V. The Transgender Law Center uses the term transgender "to describe people whose gender identity does not correspond to their birth-assigned sex and/or the stereotypes associated with that sex."[15] This definition accurately indicates that socially constructed sex stereotypes are called "gender." To distinguish the two, "sex" is a natural, biological category—male or female—whereas gender is a social category—masculine, feminine, or androgynous. Ideas about gender vary across culture, era, and race. For instance, a bourgeois male in eighteenth-century France was expected to wear lace, garters, and silk stockings whereas a bourgeois male in twenty-first-century America is not.

A gender-dysphoria diagnosis is also associated with the term "transsexual," according to the *DSM*-V. Although the following terms vary by country, in general, transsexuals have physically altered their appearance by surgical or chemical medical developments, techniques, and procedures because they have sought a treatment for their gender dysphoria. Health care sanctions and categorizes both transgender and transsexual identities through psychology and psychiatry.

Although a clinical diagnosis of gender dysphoria can lead to stigmatization, it has other benefits such as legitimizing access to medical care, use of health insurance to defray costs, having legal structures to prevent discrimination, and being able to find emotional support through friends, family, or online. The "pro-surgery" strain of medical intervention for gender dysphoria is overrepresented in medical discourse.[16] However, support for SrS is not always based on observable clinical trials that demonstrate the "success" of surgical interventions,[17] but rather social factors, such as disdain for gender ambiguity as well as political lobbying.[18]

Although both transgender and transsexual people have gender identity disorder, medical resources are only expended in the pursuit of becoming transsexual through SrS, and thus it is the focus of Green Bioethics. In order to reduce dependency on medical interventions for gender presentation, green doctors must question the diagnosis and proposed treatments for nonphysical conditions such as dysphorias, and offer more sustainable alternatives.

First and foremost, green doctors must acknowledge that medical labels should not be applied to social expectations. Gender is a subjective notion.[19] Both men and women seek cosmetic surgery to conform to highly inflected conceptions of "pure" gender. Doctors may deflect the surgical approach to masculinity and femininity by recognizing we are all transgender.[20] That is, gender is constructed by society. No one is 100 percent "feminine" or 100 percent "masculine" since people are not social caricatures. All people have pieces—even if it is the smallest bit—of both genders within them. Erasing gender fiction is environmentally sustainable. Creating gender fiction is not.

Second, it should not be the default to "cure" a "disorder" listed in the *DSM* with medical developments, techniques, and procedures when gender presentation is not a clinical matter. Many people seek SrS in order to avoid the harassment that comes from gender ambiguity. Ethicists Kristen Voigt and Harald Schmidt regard it as "clearly problematic when people resort to surgical procedures to escape stigma, bias, and discrimination."[21] While the preceding reference was to gastric banding for the obese and overweight, the argument can be applied to a variety of other settings where medical intervention is sought to avoid social problems. Voigt and Schmidt continue, "It would be far preferable to address stigma and discrimination directly, rather than have people undergo surgical procedures."[22] Gender presentation is something outside the purview of health care. Instead of substantial technological investments, the non-technological solution is acceptance of variations in human gender expression. Doctors must assure their patients that they are accepted and valued as human beings when they are gender-nonconforming.

Third, even if medical interventions are strongly desired by an individual, it should not be assumed that high-tech interventions in gender presentation are the best standard of care. Sex reassignment surgery does not have a significant record of sexual function or "social passing"[23]—that is, being viewed and accepted by others as a man or woman. Moreover, use of the health-care industry can be physically harmful when "cures" and "treatments" are overprescribed. Doctors must consider how iatrogenic disease causes physical pain and other medical externalities.

In the case of SrS, harmful side effects may include surgery to repair prolapsing, or reopen artificially created vaginas that are attempting to heal from "constructive" surgery, the carcinogenic effects of female hormones, unpredictable outbursts of violence from male hormones, recuperation from elective mastectomies, infections, or prescription of antidepressants for disappointment with surgical outcomes of "gender reassignment."[24] In one case, dissatisfaction with SrS led to a petition for euthanasia. The request was granted in 2013.[25]

An additional layer of caution is needed when SrS is anticipated for minors, and chemical prescriptions are offered along these lines. Some people within the medical industry have advocated that children who display gender dysphoria be subjected to medical developments, techniques, and procedures such as "puberty suppression" via injections of male or female hormones so that they can avoid bullying and make the "inevitable" surgical sex transition easier later in life.[26] But high-tech, prolonged, extensive, or invasive interventions do not necessarily result in best patient outcomes, particularly when they are irreversible. Simplicity can prevent the ramifications of overtreatment by evaluating the legitimacy of the *DSM* diagnosis and proposed treatments.

Fourth and finally, gradation can still be offered for those who feel compelled to present a highly specified form of gender. One gradational approach to gender presentation is nonsurgical feminization or masculinization. This can be done in a number of ways, such as altering socially determined gender markers like hair length, dress codes, hobbies, and activities. People who wish to be perceived as the other sex can modify their referencing pronouns, manner of speaking, or choice of romantic partner. Nonsurgical options for gender presentation are widely available and can be achieved simply.

Providing the patient-client any medical services they request will not necessarily result in better health outcomes or higher levels of personal satisfaction. This is especially pertinent when people are offered medical treatments for physical appearance. Sex reassignment surgery as a "routine treatment" for gender dysphoria must be reevaluated. Recognizing variety in gender presentation provides the best social environment for the person. At the same time, the natural environment is preserved when gender presentation is not viewed as a problem to be cured by medical developments, techniques, and procedures. When doctors address ailments that are clinically present, they fulfill the goals of medicine. But when treatments are given with no medical advantage to the patient, resources are wasted.

Ethical Economics: Humanistic Health Care Instead of Financial Profit

Doctors in the United States are paid extremely well, but not all physicians attend to the humanistic health care of patients. Doctors are given financial incentives to overprescribe or overtreat. "Many people believe that best practice medicine is linked to high technology, sophisticated and expensive exams. Thus, this scenario pressures mainly private physicians to work on the side of money instead of efficiency and rational and evidence-based protocols."[27] It may be tempting to acquiesce to a system that commodifies health care, yet physicians can work towards application of the fourth principle of Green Bioethics by focusing on humanistic health care. This will eliminate medical waste by rejecting unnecessary profit from services that do not benefit the somatic well-being of the person.

Doctors are incentivized to go beyond best practices or adequate treatment, and are compensated for "extra charges, unauthorized charges, [unnecessary] hospitalizations, questionable procedures, [and] unnecessary tests."[28] Some doctors choose to go into practices that are, by nature, high-profit endeavors. Doctors should reject the provision of medical developments, techniques, and procedures that fall under these categories. Cosmetic dentistry, fertility services, and non-reconstructive plastic surgery are among specialties that do not work towards humanistic health care for all people.

To be sure, there can be overlap between humanistic and financially driven treatments in the same clinical office. For instance, a dentist may offer teeth alignment so that a person can chew food, but also offer cosmetic teeth bleaching. The American Academy of Implant Dentistry reports that 10 percent of all dental clinics offer (cosmetic) dental implants.[29] Their mixed offerings make it difficult to determine if a particular doctor's office addresses humanistic health care or operates on a financial-profit model.

Moreover, there is a lack of reporting and regulation in businesses like the fertility industry. The Centers for Disease Control (CDC) recorded 464 fertility clinics in the United States during 2015, but only included clinics that self-reported in vitro fertilization (IVF) cycles and successes.[30] Fertility clinics are not required to report the number of IVF cycles performed to the CDC. In fact, fertility clinics are not even required to report their existence to the CDC.[31] This translates to a number of non-reporting fertility clinics that have very low success rates and do not publicize their information. Fertility clinics use an enormous amount of resources and have a high financial return for clinicians, but their services rarely attend to clinical disease.

Finally, some cosmetic organizations report membership numbers, but these might be lower than actual practicing aesthetic doctors. The American Society for Aesthetic Plastic Surgery distributes over 21,000 surveys to board-certified physicians practicing plastic surgery, otolaryngology, and dermatology. However, these surveys do not reflect total membership in the society or all plastic surgeons nationwide.[32] There are many more doctors that perform elective procedures not included in this organization, as well as doctors who are not members of this group who offer both reconstructive and cosmetic surgery.

To counterbalance these obstacles, doctors deciding on a particular branch of medicine to practice can select clinics and facilities that only provide humanistic treatments; choose to enter practices that are accountable to regulations and oversight to ensure that their clients will receive treatments that obtain desired outcomes; and voluntarily report their actual practices. Identifying for-profit branches of medicine are not necessarily self-evident for several reasons, but discernment will lead doctors to ethical economics.

When doctors trained in medicine provide lucrative but nonessential services, excess consumption reigns to the detriment of sustainability. Rubens Costa-Filho writes,

> Many physicians feel it is not their job to question the benefit of care, instead using a consumer satisfaction standard: "They want everything done, and it's not for me to question their motivations" . . . [But] no one is looking too closely at how much this treatment will actually benefit the patient.[33]

Cosmetic dentistry, fertility clinics, and aesthetic medicine are financially lucrative services in high demand that are not sustainable. But primary care, clinically proven treatments, effective health-care delivery, and cost-effective therapies support ethical economics by placing humanistic health care before financial profit.

Conclusion

Climate change is an issue that cuts across people, countries, and health care. Medical waste endangers the environment and the future of health care. "Current levels of consumption may challenge our ability to provide health care for future generations."[34] Rather than acceding to the current atmosphere of medical consumerism, Green Bioethics provides four principles that support doctors implementing sustainable health care. There are multiple ways doctors can participate in the principles of Green Bioethics. Working in medical deserts, halting futile treatments, offering gradational approaches to medical interventions, and avoiding for-profit elective services are examples specific to green doctors.

CHAPTER 9

The Green Health-Care Plan

oth governmental and private health-care plans act as guardians and distributors of medical resources, thus having the potential to take significant roles in sustainable health care. Large-scale insurance companies and government health-care plans will be leaders in sustainable health care when they work within the parameters of Green Bioethics. They have the necessary power and influence to shape the very fabric of sustainable health care when the principles of Green Bioethics are institutionalized.

Here, the term "health-care plans" describes both public—typically governmental—and private health insurance. These health-care plans may be purchased by an individual or family, provided by employers, or provided by governments. Within the health-care plan marketplace, both governments and employers offer totally or partially subsidized health-care plans, or unsubsidized plans. Within these plans there are typically "levels" of coverage, with various co-payment obligations, and further options such as "health savings accounts" and "flexible savings accounts" for additional fees. The diversity and complexity of health-care insurance, in addition to the variations in national health-care systems, makes the term "health-care plans" a reasonable way to describe the money, policies, choices, and obligations of health-care provision that is purchased in a market economy or offered by governments, or a combination of both.

The four principles of Green Bioethics can assist in crafting and maintaining sustainable health-care plans. This can occur when, first, health-care plans offer medical developments, techniques, and procedures geared towards basic medical resources, based on lists of prioritization. Distributive justice reduces special-interest access in health-care plans. Second, health-care plans can develop policies that are explicitly conservationist, implicitly conservationist, or a mixture of both. Resource conservation will occur when health-care needs are easily accessible and health-care wants confront logistical obstacles. Third, health-care plans can reduce dependence on medical interventions by promoting policies that prevent disease. Simplicity, furthermore, takes a gradational approach to health-care delivery when medical interventions are absolutely necessary. Fourth, ethical economics occurs when health-care plans minimize ties to the luxury pharmaceutical industry. Environmental pollution is diminished when health-care plans only use the pharmaceutical industry for drugs that support humanistic health care. Institutions are not exempt from ecological stewardship, especially when they oversee national resources. The principles of Green Bioethics will shape the future of sustainable health-care plans.

Distributive Justice: Allocate Basic Medical Resources before Special-Interest Access

Currently, health-care plans give the appearance of "free" medical procedures. It is thus more difficult to discriminate between medical resources that are generally accessible and those that are special-interest. Identifying and then suggesting that health-care plans remove basic medical resources would be monstrous, but ceasing to offer certain unnecessary, desired, superfluous treatments for well-off humans is not unethical. This limitation of special-interest access through prioritization is justifiable under certain circumstances. For instance, "limitations on the rights of individuals in the face of public health threats are firmly supported by legal tradition and ethics. All legal systems, as well as international human rights, permit governments to infringe on personal liberty to prevent a significant risk to the public," note Ronald Bayer and Amy Fairchild.[1] The resource use of health care threatens public health through climate change–related health hazards.

To be sure, access to basic health care should not be endangered when health-care plans apply the first principle of Green Bioethics. Indeed, the "soft" measures of distributive justice may prevent draconian policies in health-care plans in the future, should the environmental crisis continue on its current trajectory. This should be infinitely more palatable for policymakers than making hard decisions about prioritizing basic medical resources related to lifesaving medicine; it is easier to deny elective knee replacements than contraception. At this time, there is no reason to believe that *rationing* of basic medical resources is either desirable or necessary for environmental sustainability. However, special-interest access can justifiably be prioritized well below needs.

The difference between prioritization and rationing, as Thomas Kerz explains, is that "prioritization should lead to a definition of procedures that have either a high or a low priority." In contrast, rationing "would come into effect when prioritization is no longer an option because of even more scarce financial resources."[2] In Green Bioethics, prioritization originates not from scarce financial resources, but from scarce natural resources. Kerz continues, "If resources are scarce, medically and democratically legitimated prioritization lists should be available from which doctors could decide which procedures to perform."[3]

In public-health literature, a number of lists identifying prioritized medical resources have already been proposed.[4] Democratic engagement is an ideal and has more cachet in the United States than in other countries. It is therefore more persuasive for decentralized, individualistic nations.[5] However, other health-care plans that are determined by governments need only engage their policymakers to draft environmentally sound, and just, prioritization lists. The task of identifying basic medical resources will likely be more effective when policymakers determine medical offerings. The United Kingdom's National Health Service indicates that "reduction should be a regular item on Board agendas."[6] Since medical developments, techniques, and procedures require resources, governmental oversight offers both the rationale and the capacity to identify and implement lists of prioritizing for their health-care plans.

Health-care plans that provide special-interest access to medically unnecessary procedures like Botox and breast pumps go beyond allocation of basic medical resources.[7] Mobile catheters for adults, Accutane (isotretinoin) for acne, "female Viagra,"[8] elective surgeries to "correct" snoring, steroids for cosmetic eye cysts, Taltz (ixekizumab) injections for psoriasis, and other medical developments, techniques, and procedures identified in this book are among the items that could be deprioritized in green health-care plans. These might be still available outside of health-care plans, if they are consonant with sustainability.

It is not the obligation of health-care plans to cover special-interest access services. Prioritization is not only necessary, it is logical. Indeed, "it is difficult to imagine a health care plan that will achieve all the desired goals without some form of creative prioritization," observes David Crippen.[9] Health-care plans can support sustainability by prioritization of basic medical resources before special-interest access.

Resource Conservation: Provide Health-Care Needs before Health-Care Wants

Authors writing for the *Lancet* have proposed that "An integrated approach to attempting to reduce the adverse effects of climate change requires at least three levels of action. First, policies must be adopted to reduce carbon emissions, and thereby slow down global warming and eventually stabilize temperatures. Second, action must be taken on the links connecting climate change and adverse health. Third, appropriate public health systems should be put into place to deal with adverse outcomes."[10] Green health-care plans are suited for the first level: policy.

Policies add structure to values, which are then disseminated into social, political, and medical environments. Policies are pragmatic and efficient since individuals and groups do not need to evaluate alternatives each time a choice is presented. Policies also make decision making about complex issues like health care and the environment easier since issues have, ideally, been deliberated, debated, and assessed prior to establishment. Health-care plans can utilize the second principle of Green Bioethics by providing health-care needs before health-care wants through policies that are explicitly conservationist, implicitly conservationist, or a mixture of both.

Currently, governments that sponsor health-care plans conserve resources through a range of tactics. In surveying health-care plans across the world, some strategies rely on explicitly conservationist measures to curtail minimally beneficial medical offerings to patients. Such is the case in New Zealand, where "limiting (and withdrawing) intensive therapies are common practices . . . and are usually well accepted by other health professionals, patients and their families."[11] Conservationist measures in health-care plans are embedded in New Zealand's culture. Although New Zealand does this for financial and not ecological reasons, oftentimes the two overlap. Conserved health-care resources limit environmental waste and save money. It should be reiterated that the withdrawal of non-beneficial treatment is not medically necessary by definition, and therefore a health-care want. Other national health-care plans place restrictions on nonurgent medical developments, techniques, and procedures, conserving resources through logistical barriers.[12]

The Australian government provides health-care plans for its citizens, augmented by private health-care plans. Under public, governmental health-care plans, implicit prioritization of nonessential medical developments, techniques, and procedures is implemented by bureaucratization. Writing from Australia, Ian M. Seppelt records, "For elective admissions (mainly elective surgery) rationing is achieved by waiting, at times up to a year for non-urgent surgery, and patients are treated by a hospital-appointed doctor, rather than necessarily the doctor of their choice."[13] These strategies minimize burdens on health care, make access to health-care needs more widely available, and place health-care wants on a lower level. The Australian model has made inroads into conservation while still allowing for health-care wants to be allocated after a waiting period. A more effective and successful approach to resource conservation would eliminate elective treatments entirely. Finally, some countries could utilize multilevel health-care plans to curtail excessive medical waste by combining explicit and implicit conservationist measures.

In Germany, the matter of health-care distribution has been evaluated, and strategic measures have been proposed. German policymakers have suggested that the health-care plans prioritize health-care needs and reduce excess health-care spending through a tiered system. In this scenario,

> The first tier or basic level will probably be a compulsory insurance, covering life-threatening and acute diseases. The second-tier will demand more copayments and will leave extent of coverage and premiums to every insured. The third-tier will round off coverage for even marginal health problems.[14]

Health-care plans in Germany can reduce medical waste effectively through targeted adjustments at many levels, recognizing that "saying no to infinite health care demands will be achieved by a mix of limitations set by the organization of the health care system itself and by personal preferences."[15] Limitations and specifics of green health-care plans and sustainability preferences will vary by country.

It is significant that in the United States, Tom Beauchamp and James Childress proposed a multi-tiered approach to health-care plans, whereby basic levels of care include "public health measures and preventive care, primary care, acute care, and special social services for those with disabilities."[16] These health-care needs are clearly aligned with the goals of medicine and provide quality of life and function. Beauchamp and Childress's proposed second and third levels would accept medical developments, techniques, or procedures that are beyond "basic care." Health-care wants go beyond basic care and offer standard of living and enhancement. Beauchamp and Childress's tiers could translate into health-care plan policies in the United States.

Indeed, the Affordable Care Act partially adopted a layered approach to health-care plans; however, there were still lacunae in public-health measures, while nonnecessities were covered under even the lowest-level health-care plan. For example, the most basic health-care plan of Blue Cross and Blue Shield in North Carolina, called the "Blue Advantage Bronze Plan," is free for qualified low-income residents and covers bariatric surgery for morbid obesity, thirty visits a year for chiropractic services, infertility services including drugs and three medical ovulation induction cycles per lifetime per member, and private duty nursing.[17] A tiered approach to resource conservation in health-care plans provides the most flexibility and freedom, but ultimately lacks the rigor to address the current environmental crisis by focusing on unfettered choice rather than ecology.

In every sector of life, ecologists recognize a limited number of resources. Each person competes for these resources, and some are left without any health-care needs met.[18] Resource conservation must drive health-care plans, thus directing policies towards honest identification, and provision of, health-care needs.

Simplicity: Reduce Dependence on Medical Interventions

Both private and governmental health-care plans can apply the third principle of Green Bioethics by promoting strategies that prevent diseases or medically deleterious conditions and using a gradational approach to health-care delivery when medical interventions are absolutely necessary. Simplicity is particularly relevant as health-care plans address diseases and conditions associated with being overweight.[19]

Approximately 70 percent of the United States population are obese or overweight.[20] This translates to health-care plans being overburdened with people whose health is failing due to what is often a lifestyle choice. The United States alone "spends $147 billion each year to treat obesity, $116 billion more to treat direct costs of diabetes, and hundreds of billions more to treat cardiovascular disease and cancer that many suspect are related to the Western diet."[21] Current health-care plan-sponsored treatments for obesity-related conditions include medications like metformin for type 2 diabetes, gastric bypass surgery, heart surgery, stents, pacemakers, assisted reproductive technologies, positive airway pressure (PAP) machines for sleep apnea,[22] and Viagra for obesity-related impotence.[23] These medical developments, techniques, and procedures intervene at a higher level than necessary since they mask the "problem" of obesity but do not fix the underlying condition. Yet, they are covered under health-care plans since physicians, patients, and the public—who are generally overweight—consider lifestyle changes "too much work." Their lobby has influenced health-care plans to the extent that policies seem to agree that it is more convenient to prescribe a pill or undergo surgery than consistently work for change.[24]

Instead of applying a simple approach to obesity, health-care plans are inventing ever more complicated and resource-intensive ways to address personal choices that have escalated to require medical interventions. Furthermore, some health-care plans are considering expanding their coverage of treatments related to the preventable "disease" of obesity. In 2014, the British National Institute for Health and Care Excellence (NICE) suggested that the UK's National Health Service (NHS) cover bariatric surgery to reverse type 2 diabetes. NICE alleged that this medical intervention might also be concomitant for voluntary joint-replacement surgery as a result of obesity.[25] If the NHS adopts this proposal, the resource use of health care will increase, affecting climate change–related health-care costs as atmospheric pollution is released.[26]

Simplicity seeks to align the health of individuals with the health of the planet; thus preventative strategies must be starting points for health-care plans. Instead of health-care plans covering pills, surgery, and devices, prevention and gradation should be employed. Health-care plans can support obesity prevention in many ways, such as financing bonuses to maintain a healthy body mass index (BMI), or other incentives for weight loss.

If prevention is unappealing to health-care plans, a gradational approach would reduce resource use. A first-level solution for weight-related conditions would support lifestyle changes. Health-care plans could reimburse doctors for their time discussing exercise programs, nutrition, and caloric values with patients. Other health-care professionals could work with individuals to make gradual, sustained lifestyle adaptations, such as eating more nutrient-dense food, increasing physical activity, and altering attitudes towards food consumption.

As a second-level solution, health-insurance plans can purposefully limit the frequency and number of high-tech surgeries that are performed to reverse obesity by offering invasive surgeries on a case-by-case, rather than a minimum qualification, basis. Health-care plans must not ignore the fact that obesity—in the vast majority of cases—is simply a result of macronutrient overconsumption. It is not a medical condition, although it can cause medical conditions.[27]

In some places there are impassioned calls to "give the prevention of disease the priority that it deserves but currently lacks . . . if we want to avoid a situation where more than half of the population is taking carbon intensive drugs."[28] Given the impact on shared resources, and the continual demand on health-care plans to address this largely preventable condition, simple and ecological tactics in "chasing the elusive [obesity] epidemic" must be implemented.[29] Obesity will remain a worldwide issue.[30] Health-care plans should support prevention of obesity, only later followed by gradation in the treatment of obesity.

Ethical Economics: Humanistic Health Care Instead of Financial Profit

Pharmaceutical corporations are powerful forces, lobbying for novel drugs— some of them with minimal benefit—offering doctors incentives for prescribing their version of a pill, and requiring an enormous amount of resources. A large section of the pharmaceutical industry dedicates itself to marketing drugs that do not cure or treat clinical diseases. This bolsters profits while also causing a strain on the environment. Health-care plans must prioritize humanistic health care by minimizing ties to the luxury pharmaceutical industry.[31] Eliminating drugs, like Viagra, that are driven by profit, not humanistic health care is one application of ethical economics.

Viagra (sildenafil citrate) is an incredibly lucrative pharmaceutical that is heavily marketed to treat the symptoms of male "erectile dysfunction."[32] This is not a universal conception, neither does it threaten lifespan or cause mortality. Medical labels, although presented as pejorative, represent an ordinary part of male sexual responses and normal reactions to excitement, stress, fatigue, or diet. Yet, direct market-to-consumer advertisements for Viagra to middle- and upper-class clients leads to patients making demands on health-care plans for a pharmaceutical that is not clinically necessary, but rather offered as a means for "sexual self-determinism for privileged men."[33]

While the range of male sexual response is not a disease, it can point to underlying clinical conditions that can, and should, be addressed. "Impotence" can be caused by medical problems such as "diabetes, heart disease, prostate surgery and spinal cord injury. It also can be psychological or [due to] a drug side effect."[34] However, it makes little sense to prescribe a pill to treat the *effects* of a condition rather than the condition itself. For instance, the Department of Defense spent $84.24 million dollars on erectile dysfunction prescriptions in 2014. Over half of the prescriptions were given for psychological, not physical, reasons.[35] Health-care plans must address the disease, if there is one, instead of innocuous symptoms. The supply of Viagra cannot be considered a legitimate fulfillment of health-care plans since insurance is not meant to be a lifestyle manager.

There are numerous compelling reasons for minimizing ties with the luxury pharmaceutical industry, such as financial exploitation, environmental waste, and drug dependence. Pharmaceuticals are the second largest contributor to carbon emissions within health care. Although ecological sensibilities, or "corporate social responsibility," are unlikely to form in the cultures of drug companies themselves, health-care plans must consider the environmental ramifications of overprescription. Luxury pharmaceuticals—such as Viagra— are an enormous source of resource use and need not be covered under health-care plans.

At the same time, health-care plans should cover medically necessary pharmaceuticals. Certain drugs such as antiretrovirals, antibiotics when appropriate, birth control pills, and other forms of humanistic health care should be used, distributed, and available at fair cost. Governments that provide health-care plans should be particularly interested in humanistic health care since it aligns with their prima facie goals of basic care for all citizens and national interests in a stable ecosystem.

Conclusion

Current health-care plans are antithetical towards sustainable health care because there is little sense of planetary limitation and immense pressure from medical consumers to have continually expanded access to the health-care industry. "Because people in modern societies expect the state to defray all or part of the cost of what is deemed a 'medical service,' where we draw the line between 'health care' and 'not health care' is informed more by economics and political considerations than by medical or scientific judgments."[36] Some health-care plans in the United States fund ancillary services such as acupuncture, massage therapy, and "alternative medicine disciplines."[37] However, these are offered not as low-resource alternatives to traditional health care, but in addition to standard health care. Supplementary or "bonus" offerings make particular health-care plans more appealing to insurance shoppers, clearly indicating the market-driven consumerism. Society must move away from this mentality and back towards health-care plans as safety nets, not commodities.

Recognizing the limits of resources and burdens of consumer demand, the NHS recommends that "every organisation needs to consider their approach to commissioning, sourcing and buying. This will include considering if it needs to be purchased in the first place."[38] The onus of responsibility is on insurance companies and governments to restructure health-care plans to cover essential medical developments, techniques, and procedures in the least environmentally damaging way possible. Utilizing the principles of Green Bioethics will guide health-care plans towards essential care for all individuals, while also protecting the planet, and thus the health, of everyone.

Conclusion

Ecological sustainability is among the most significant ethical situations facing not only humanity, but also health care. Consequently, Green Bioethics provides a coherent environmental framework for modern health care by proactively surveying the sustainability of medical developments, techniques, and procedures. Yet, Green Bioethics will confront challenges from conventional bioethics, which firmly retains its biomedical sensibilities, prioritizing the physician-patient relationship and the four principles of biomedical ethics.

Since a prominent feature of traditional biomedical ethics is the physician-patient relationship, Green Bioethics must be relevant to, and conversant with, the language and principles of Western bioethics. In order to demonstrate opportunities for rapprochement of ecology and biomedical ethics, one clinical practice and one area for further work for sustainability in health care—both centered on the doctor-patient relationship—will be identified. These situations will speak directly to clinicians and other health-care practitioners who abide by the conventional biomedical ethics.

As a practice, "green informed consent" would combine best patient care and sustainability. The environmental impact of medical developments, techniques, and procedures would then be explained to patients when making medical decisions. As an area for further work to be done, the scope of health-care concern within the physician-patient relationship could be expanded from the one to the many. With a corporate notion of "the patient," doctors would be expected to adjudicate between resource-intensive treatments for the patient in front of her and resource impact of these treatments on all potential patients. The reunion of ecology and bioethics, and the union of environmental sustainability and traditional biomedical ethics can be accomplished within Green Bioethics.

Green Informed Consent

Informed consent is a decision-making process that follows a medical diagnosis. Informed consent is a standard practice of biomedical ethics. It is based on the principle of respect for autonomy, which maintains that capable adults are able to make their own medical decisions when they have proper information. After a diagnosis, patients are educated about medical options and alternatives to treatments, with the benefits and drawbacks of each. They may ask clarifying questions related to time of recovery, cost of proposed treatments, length of treatments, and side effects, for example, and can also voice their values. The choices that patients make, if clinically indicated and within the competencies of the physician, should be respected and therefore carried out, according to traditional Western biomedical ethics. Informed consent is a part of health care in everything from cancer to contraception. Since this practice is common, physicians, nurses, and medical assistants can offer "green informed consent" as a practice of Green Bioethics.

Green informed consent would provide the patient with information on the estimated resource expenditure of the respective treatments, along with the resource use of any anticipated side effects or follow-up care. Notice that green informed consent would mirror both the information about the variety of treatments to choose from—along with resource-use data—and information about the ramifications of each choice, from recovery time to cost, to resource use.

While some health-care providers might be concerned that green informed consent would compromise best treatment, it would not necessarily be the case that the best treatments would be the most resource-intensive. And, even if the best treatments were the most resource-intensive, it would not necessarily be the case that ecologically minded patients would choose the least impactful medical treatment. In life, many people make decisions by balancing values. Just as carbon offsets are offered to people using air travel, ecology can be creatively integrated into a medical plan.

If we imagine the application of green informed consent for a forty-two-year-old female diagnosed with breast cancer, it might be that chemotherapy would be more resource-intensive than a mastectomy. Yet, because a patient values her breasts, or dislikes surgery, she might choose to "offset" the high-resource procedure through other tradeoffs in her life. The patient might use a more fuel-efficient vehicle, eliminate red meat from her diet, or limit medical treatments in the future. But, she can only readjust her life plans if she has full and appropriate information about the resource impact of her choices.

Patients are permitted to make medical decisions based on their own preferences, goals, and desires. Green informed consent would respect the values of patients who endorse a conservationist lifestyle. Currently, health-care workers may tailor information about medical treatments to the individual's values, supporting the ideals of the patient and community. Making reference to the projected amount of resources that will be used in the informed-consent process would give patients the fullest information available, leading to a truly informed decision. In fact, one could argue that once resource expenditure of a given procedure is known, a doctor would be withholding relevant information if she did not provide the patient with environmental data. One article from a 2009 issue of the American Medical Association's ethics journal indicated that, in some cases, educating patients on the environmental effect of various procedures is already included in the informed consent in some health-care facilities.[1]

Of course, informed consent is a two-way street. Doctors will have to be convinced that ecological aspects of health care are salient factors in one's decisions about medical treatments. In order for this to be effective, the doctor must be able to determine the resource use of a medical procedure. Thus, this first practice of Green Bioethics requires a second component.

Implementing green informed consent would require examining resource use of individual medical developments, techniques, and procedures. Once these medical developments, techniques, or procedures were identified with an amount of resources, doctors could refer to the information at patient consultations. This information should also be accessible to the public as a part of informed decision making and medical consumer education. If there were a revolution in health care, these environmental valences might even factor into wider democratic deliberations on which procedures health-care facilities and medical insurance companies are willing to offer.

With growing consensus that the balance of the ecosystem mandates attention and action, and with green hospital practices already abounding in health care—from recycling bins in clinics to local food in cafeterias—green informed consent is a natural progression from institutions to individuals. Of course, suggesting that resource use be factored into health-care decisions leads some people to believe that the doctor-patient relationship would be imperiled by environmental interference.

Both the mention of resource use of medical procedures in the United States and proposed policies about distribution of sustainable medical procedures are ignored or strongly rebuffed.[2] It might be feared that green informed consent, which includes environmental considerations, might imperil "the physician/patient relationship, which is built upon trust. If a patient must wonder whether their doctor is worried more about rationing resources or saving the environment, then the physician/patient relationship is damaged."[3] Under Beauchamp and Childress's principles of biomedical ethics, the physician's responsibility is to the patient above all else. Doctors in the United States, who do not have to work within the parameters of socialized health care and who have a wide assortment of medical developments, techniques, or procedures to offer to their patients, do not consider the ecological effects of health-care options on the larger population. Health-care providers may feel that their ethical obligations extend only to the patient in front of them, and best patient care might include numerous resource-intensive interventions, tests, or treatments. However, there is a way to deliberate between individual patient needs, resource limitation, and planetary health in Green Bioethics.

The Corporate Patient

Green Bioethics will need to continue to address sustainable health care from a Western, atomistic perspective. As an area for further work, it must be argued that patient-centered care includes all potential patients, including those outside of the hospital. Enlarging the conception of "the patient" can alleviate the seeming tension between the obligations of the physician to her patient, the rights of all individuals to health care, and the reality of resource depletion.

It is well documented that climate change causes health problems.[4] There are many ill and ailing people who need medical attention but cannot access any health care. These people are impacted by the choices that doctors make inside of health-care facilities, even though they may be unable to avail themselves of treatments. Thus, vulnerable populations face a double burden of not being able to receive medical care and being affected by climate-change health hazards that are caused by the resource use and carbon emissions of health-care facilities. In fact, even the individual patient who is submitting to medical treatments is impacted by the pollution of hospitals and health-care facilities upon discharge. Physicians must be concerned with the health of their patients when they are undergoing treatment as well as when they are not, that is, in day-to-day life.

Physicians work against themselves when they view the patient in front of them as a monad without considering the needs of all their patients—present and future. The resources required to provide medical developments, techniques, and procedures within the hospital affect those outside of it. Sustainable medical treatments for individual patients will have the aggregate effect of preventing thousands of sick and dying patients from requiring medical care. The physician-patient relationship must be thought of as not one-to-one, but one-to-many. In this way, appeals to the corporate patient benefit from a public-health approach.

While "medical ethics is more concerned with individual autonomy and the duties of single health professionals, public health ethics focuses more on equity and efficiency in the distribution of health resources as well as on the community in having its health protected."[5] In public health, the rights and freedoms of individual patients are important, but they are not the only factor to consider in medical care and treatment. Public health expands the values of Western biomedical ethics to all individuals affected by medical actions. The integration of public-health initiatives into the traditional principles of biomedical ethics provides precedent for those who argue that there are ethical obligations beyond a specific doctor-patient relationship. The innovation of public health as a concept also demonstrates the continual need to reevaluate and integrate the demands of a rapidly changing world for health care. The work ahead of Green Bioethics will require a global commitment.

Cheryl C. Macpherson and Muge Akpinar-Elci contend that "bioethicists, like other professionals, have opportunity and influence with which to alter emissions-related policies."[6] This book has presented four principles to determine the sustainability of medical developments, techniques, and procedures, thus upending traditional approaches to biomedical ethics and environmental bioethics. It has offered multilevel suggestions for green individuals, doctors, and health-care plans, which give ownership to each person or entity involved in health care. Finally, it has proposed green informed consent and the notion of the corporate patient as opportunities for rapprochement of ecology and biomedical ethics. Henceforth, ecologists, ethicists, clinicians, and bioethicists must emerge and leverage their collective influence—and intelligence—to advocate practices and policies that support sustainable health care utilizing the principles of Green Bioethics.

Notes

INTRODUCTION

1. Bruno Martini, "The Anthropocene: Humankind as a Turning Point for Earth," *Astrobiology*, June 24, 2013.

2. Furthermore, "the 'high-variant' projection, which assumes an extra half of a child per woman (on average) compared to the medium variant, implies a world population of 10.9 billion in 2050 and 16.6 billion in 2100. The 'low-variant' projection, where women have half a child less, on average, than under the medium variant, would produce a population of 8.3 billion in 2050." Notably, "a constant difference of only half a child above or below the medium variant would result in a global population in 2050 of around 1.3 billion more or less compared to the medium variant of 9.6 billion." United Nations Division of Economic and Social Affairs, *World Population Prospects: The 2012 Revision*, vol. 1, *Comprehensive Tables* (New York: United Nations, 2013), xv, xviii.

3. Corey J. A. Bradshaw and Barry W. Brook, "Human Population Reduction Is Not a Quick Fix for Environmental Problems," *PNAS: Proceedings of the National Academy of Sciences of the United States of America* (October 27, 2014): 16613.

4. Peter H. Gleick and Heather S. Cooley, "Energy Implications of Bottled Water,"

Environmental Research Letters 4, no. 1 (2009): 014009.

5. Pacific Institute, "Fact Sheet: Bottled Water and Energy: Getting to 17 Million Barrels," December 2007, p. 2, https://pacinst.org.

6. "CO$_2$ Time Series 1990–2015 per Region/Country," European Commission, Emissions Database for Global Atmospheric Research, http://edgar.jrc.ec.europa.eu.

7. "CO$_2$ Time Series 1990–2015 per Region/Country."

8. Julia Whitty, "Diagnosing Health Care's Carbon Footprint," *Mother Jones*, November 10, 2009.

9. Jeanette W. Chung and David O. Meltzer, "Estimate of the Carbon Footprint of the US Health Care Sector," *Journal of the American Medical Association* 302, no. 18 (2009): 1970–72.

10. Ian Roberts, "The NHS Carbon Reduction Strategy," *BMJ* 38, no. 7689 (2009): 248.

11. World Health Organization, *Global Health Risks: Mortality and Burden of Diseases Attributable to Selected Major Risks* (Geneva: WHO Press, 2009), 24.

12. Kim Knowlton et al., "Six Climate Change–Related Events in the United States Accounted for About $14 Billion in Lost Lives and Health Costs," *Health Affairs* 30, no. 11 (2011): 2167–76.

13. Anthony Costello et al., "Managing the Health Effects of Climate Change," *Lancet* 373, no. 9676 (2009): 1693–733.

CHAPTER 1. ENVIRONMENTAL BIOETHICS

1. Fritz Jahr and Hans-Martin Sass, "Bio-Ethics—Reviewing the Ethical Relations of Humans towards Animals and Plants," *JAHR-European Journal of Bioethics* 1, no. 2 (2010): 227.

2. Jahr and Sass, "Bio-Ethics," 230.

3. Van Rensselaer Potter, *Global Bioethics: Building on the Leopold Legacy* (East Lansing: Michigan State University Press, 1988), 2; Van Rensselaer Potter, "Bioethics: The Science of Survival," *Perspectives in Biology and Medicine* 14, no. 1 (1982): 127–53.

4. Warren T. Reich, ed., *The Encyclopedia of Bioethics*, vol. 1 (New York: Macmillan, 1978), xix.

5. Tom Beauchamp and James Childress, *Principles of Biomedical Ethics*, 1st ed. (New York: Oxford University Press, 1979).

6. Warren T. Reich, "The Word 'Bioethics': The Struggle Over Its Earliest Meanings," *Kennedy Institute of Ethics Journal* 5, no. 1 (1995): 19–34.

7. Cristina Richie, "A Brief History of Environmental Bioethics," *Virtual Mentor: American Medical Association Journal of Ethics* 16, no. 9 (2014): 749–52.

8. New York University School of Medicine, "MD/MA in Bioethics," https://med.nyu.edu/education. See also *Journal of Medical Humanities* 23, no. 1 (2002): 3–92; *Health Affairs* 30, no. 5 (2011): 810–997; and *Health Care Ethics USA* 23, no. 3 (2015): 1–41. American Society for Bioethics and Humanities, National Undergraduate Bioethics Conference, Georgetown University, April 5–7, 2013.

9. David Resnik, *Environmental Health Ethics* (Cambridge: Cambridge University Press, 2012).

10. Van Rensselaer Potter, *Bioethics: Bridge to the Future* (New Jersey: Prentice-Hall, 1971); Jessica Pierce and Andrew Jameton, *The Ethics of Environmentally Responsible Health Care* (New York: Oxford University Press, 2004).

11. Health Care Without Harm, *Healthier Hospitals Initiative* (Reston, VA: Health Care Without Harm, n.d.), 3.

12. Health Care Without Harm, *Healthier Hospitals Initiative*, 3.

13. Di(2-ethylhexyl)phthalate (DEHP)/Polyvinyl Chloride (PVC) is found in a multitude of hospital devices such as breast pumps, enteral nutrition products, parenteral infusion devices and sets, general urological products, exam gloves, umbilical vessel catheters, and vascular catheters. In 2001, the Food and Drug Administration (FDA) assessed DEHP/PVC exposure in hospital settings and found these medical devices to be toxic. Department of Health and Human Services, *Public Health Notification: PVC Devices Containing the Plasticizer DEHP* (Rockville, MD: Food and Drug Administration, July 12, 2002). Ten years later, the European Union's Scientific Committee on Emerging and Newly-Identified Health Risks (SCENIHR) produced a similar warning. Scientific Committee on Emerging and Newly-Identified Health Risks, *Preliminary Opinion on the Safety of Medical Devices Containing DEHP-Plasticized PVC or Other Plasticizers on Neonates and Other Groups Possibly at Risk* (Luxembourg: European Union, 2014).

14. Healthier Hospitals Initiative, "Participating Hospitals," http://healthierhospitals.org.

15. Dignity Health, "Our Mission, Vision and Values," https://www.dignityhealth.org.

16. Dignity Health, *Statement of Common Values* (San Francisco: Dignity Health, 2013), 1–3.

17. The California Climate Action Registry closed in December 2010. California Climate Action Registry, http://www.caclimateregistry.org.

18. Dignity Health, "Environment," https://www.dignityhealth.org.

19. Dignity Health, "Sustaining Our Healing Ministry: Fiscal Year 2013 Social Responsibility Report," 15, https://www.dignityhealth.org/cm/Media/documents/2010-Social-Responsibility-Report.pdf.

20. Deborah Kotz, "Hospitals Take Steps to Set Healthy Examples for Patients," *Boston Globe*, March 31, 2014.

21. Kotz, "Hospitals Take Steps."

22. United Kingdom, *Climate Change Act 2008*, ch. 27, http://www.legislation.gov.uk/ukpga/2008/27/pdfs/ukpga_20080027_en.pdf.

23. Jeanette W. Chung and David O. Meltzer, "Estimate of the Carbon Footprint of the US Health Care Sector," *Journal of the American Medical Association* 302, no. 18 (2009): 1970.

24. National Health Service Sustainable Development Unit, *Saving Carbon, Improving Health: NHS Carbon Reduction Strategy for England* (London: NHS Sustainable Development Unit, 2009).

25. National Health Service Sustainable Development Unit, *Saving Carbon*, 24.

26. National Health Service Sustainable Development Unit, *Saving Carbon*, 67.

27. National Institute for Health Research, "Carbon Reduction Guidelines" (October 2010), 4, https://www.nihr.ac.uk/research-and-impact/documents/NIHR-Carbon-Reduction-Guidelines.pdf.

28. Katy Lyle et al., "Carbon Cost of Pragmatic Randomised Controlled Trials: Retrospective Analysis of Sample of Trials," *BMJ* 339 (2009): b4187.

29. Jill Burnett et al., "Towards Sustainable Clinical Trials," *BMJ* 334 (2007): 671–73.

30. National Institute for Health Research, "Carbon Reduction Guidelines," 15.

31. National Institute for Health Research, "Carbon Reduction Guidelines," 2.

CHAPTER 2. GREEN BIOETHICS

1. Rachel Carson, *Silent Spring* (Boston: Houghton Mifflin, 1962), 20.

2. An early version of this concept was published as Cristina Richie, "Building a Framework for Green Bioethics: Integrating Ecology into the Medical Industry," *Health Care Ethics USA* 21, no. 4 (2013): 7–21.

3. Paul R. Ehrlich and John P. Holdren, "Impact of Population Growth," *Science* 171, no. 3977 (1971): 1212–17; Marian R. Chertow, "The IPAT Equation and Its Variants," *Journal of Industrial Ecology* 4, no. 4 (2000): 13–29.

4. Amos Hawley, "Ecology," in *International Encyclopedia of Population*, vol. 1, ed. John

Ross (New York: Free Press, 1982), 159–63.

5. Gro Harlem Brundtland, *Report of the World Commission on Environment and Development: "Our Common Future"* (New York: United Nations, 1987), 16.

6. James Hansen et al., "Target Atmospheric CO_2: Where Should Humanity Aim?," *Open Atmospheric Science Journal* 2 (2008): 217–31; "Trends in Atmospheric Carbon Dioxide: Recent Monthly Average Mauna Loa CO_2," U.S. Department of Commerce National Oceanic and Atmospheric Administration Earth System Research Laboratory, Global Monitoring Division, https://www.esrl.noaa.gov.

7. Brian Launder and J. Michael T. Thompson, *Geo-engineering Climate Change* (New York: Cambridge University Press, 2009); Bala Govindasamy and Ken Caldeira, "Geoengineering Earth's Radiation Balance to Mitigate CO_2-Induced Climate Change," *Geophysical Research Letters* 27, no. 14 (2000): 2141–44.

8. Michael S. Northcott, "The Concealments of Carbon Markets and the Publicity of Love in a Time of Climate Change," *International Journal of Public Theology* 4, no. 3 (2010): 303.

9. "Carbon Dioxide Emissions (CO_2), Metric Tons of CO_2 Per Capita (CDIAC)," United Nations Statistics Division Millennium Development Goals Indicators, http://data.un.org/Default.aspx.

10. The carbon emissions of a few medical procedures have been calculated. For instance, a heart bypass operation in the UK emits 1.1 tonnes of CO_2. Mike Berners-Lee, *How Bad Are Bananas? The Carbon Footprint of Everything* (London: Profile Book, 2010), 131–32. The carbon footprint for one cataract operation in Cardiff, UK, is 181.8 kg CO_2 equivalent. D. S. Morris et al., "The Carbon Footprint of Cataract Surgery," *Eye* 27 no. 4 (2013): 495–501.

11. However, see Cristina Richie, "What Would an Environmentally Sustainable Reproductive Technology Industry Look Like?," *Journal of Medical Ethics* 41, no. 5 (2015): 383–87.

12. For a consideration of bioethics outside of Western academia, see Angeles Tan Alora and Josephine M. Lumitao, eds., *Beyond a Western Bioethic: Voices from the Developing World* (Washington, DC: Georgetown University Press, 2001).

13. National Commission for the Protection of Human Subjects of Biomedical and Behavioral Research, *The Belmont Report: Ethical Principles and Guidelines for the Protection of Human Subjects of Research* (Washington, DC: US Government Printing Office, 1978).

14. Tom Beauchamp and James Childress, *Principles of Biomedical Ethics*, 5th ed. (New

York: Oxford University Press, 2001), 13.

15. Daniel Callahan, "Principlism and Communitarianism," *Journal of Medical Ethics* 29, no. 5 (2003): 287–91.

16. Jon Fuller and James Keenan, "Educating in a Time of HIV/AIDS: Learning from the Legacies of Human Rights, the Common Good, and the Works of Mercy," in *Opening Up: Speaking Out in the Church*, ed. Julian Filochowski and Peter Stanford (London: Darton, Longman & Todd, 2005), 98.

17. The contextual features include confidentiality, allocation of resources, religion, and public health. Albert R. Jonsen, Mark Siegler, and William J. Winslade, *Clinical Ethics: A Practical Approach to Ethical Decisions in Clinical Medicine*, 8th ed. (New York: McGraw-Hill, 2015), 9, 165–236.

18. Aristotle, *Politics*, trans. Ernest Barker (Oxford: Oxford University Press, 1995), I. 2, 1253 a; Aristotle, *Nicomachean Ethics*, trans. Martin Ostwald (New York: Macmillan, 1962), book 5, ch. 3. John Rawls, *Theory of Justice* (Cambridge, MA: Harvard University, 1971), 3. Friedrich Nietzsche, *On the Genealogy of Morals*, trans. Walter Kaufmann (New York: Vintage Books, 1967), 70–71. Willis Jenkins, *Ecologies of Grace: Environmental Ethics and Christian Theology* (Oxford: Oxford University Press, 2008), 63. Norman Daniels, *Am I My Parent's Keeper? An Essay on Justice between the Young and Old* (New York: Oxford University Press, 1990).

19. Beauchamp and Childress, *Principles of Biomedical Ethics*, 5th ed., 3; italics in the original.

20. Aldo Leopold, *A Sand County Almanac and Sketches Here and There* (Oxford: Oxford University Press, 1968), 224.

21. Jonsen, Siegler, and Winslade, *Clinical Ethics*, 7th ed.

22. Soren Holm, "Not Just Autonomy: The Principles of American Biomedical Ethics," *Journal of Medical Ethics* 21, no. 6 (1995): 332.

23. The Albigensians extolled suicide and condemned procreation. Albert Jonsen, *Responsibility in Modern Religious Ethics* (Washington, DC: Corpus Books, 1968), 202.

24. Compare with the Voluntary Human Extinction Movement (VHEM), www.vhemt. org. See also the Church of Euthanasia, which stands on the four pillars of "suicide, abortion, cannibalism, and sodomy" (i.e., any non-procreative sex). *Church of Euthanasia*, www.churchofeuthanasia.org.

25. Marcello Di Paola and Mirko Daniel Garasic, "The Dark Side of Sustainability: On Avoiding, Engineering, and Shortening Human Lives in the Anthropocene," *Rivista di*

Studi sulla Sostenibilità 3, no. 2 (2013): 59–81.

26. Leonardo Boff, *Cry of the Earth, Cry of the Poor* (Maryknoll: Orbis Books, 1997); James Dwyer, "How to Connect Bioethics and Environmental Ethics: Health, Sustainability, and Justice," *Bioethics* 23, no. 9 (2009): 497–502.

27. Carl Nathan and Otto Cars, "Antibiotic Resistance—Problems, Progress, and Prospects," *New England Journal of Medicine* 371, no. 19 (2014): 1761–63.

28. See Edmund D. Pellegrino and David C. Thomasma, *A Philosophical Basis of Medical Practice: Toward a Philosophy and Ethic of the Healing Professions* (New York: Oxford University Press, 1981); Edmund D. Pellegrino and David C. Thomasma, *The Virtues in Medical Practice* (New York: Oxford University Press, 1993).

CHAPTER 3. DISTRIBUTIVE JUSTICE

1. Jon Fuller and James Keenan, "Educating in a Time of HIV/AIDS: Learning from the Legacies of Human Rights, the Common Good, and the Works of Mercy," in *Opening Up: Speaking Out in the Church*, ed. Julian Filochowski and Peter Stanford (London: Darton, Longman & Todd, 2005), 98.

2. One of the most influential outlines of justice as a theory is John Rawls, *Theory of Justice* (Cambridge, MA: Harvard University Press, 1971).

3. Jessica Pierce and Andrew Jameton, "Sustainable Health Care and Emerging Ethical Responsibilities," *Canadian Medical Association Journal* 164, no. 3 (2001): 367.

4. Immanuel Kant, *On the Metaphysics of Morals and Ethics*, trans. Thomas Kingsmill Abbot (Overland Park, KS: Digireads Publishing, 2017).

5. United Church of Christ Commission on Racial Justice, *Toxic Wastes and Race in the United States: A National Report on the Racial and Socio-Economic Characteristics of Communities with Hazardous Waste Sites* (New York: United Church of Christ, 1987).

6. Robert M. Veatch, "How Many Principles for Bioethics?," in *Principles of Health Care Ethics*, 2nd ed., ed. Richard Edmund Ashcroft et al. (West Sussex, England: John Wiley & Sons, 2007), 43–50.

7. Tom Beauchamp and James Childress, *Principles of Biomedical Ethics*, 4th ed. (New York: Oxford University Press, 1994).

8. Edmund D. Pellegrino, "The Nazi Doctors and Nuremberg: Some Moral Lessons Revisited," *Annals of Internal Medicine* 127, no. 4 (1997): 307–8. Philip G. Zimbardo, "On the Ethics of Intervention in Human Psychological Research: With Special Reference to the Stanford Prison Experiment," *Cognition* 2, no. 2 (1973): 243–56.

David Satcher, "The Legacy of the Syphilis Study at Tuskegee in African American Men on Health Care Reform Fifteen Years after President Clinton's Apology," *Ethics and Behavior* 22, no. 6 (2012): 486–88. Gretchen Reynolds, "The Stuttering Doctor's 'Monster Study,'" in *Ethics: A Case Study from Fluency*, ed. Robert Goldfarb (San Diego, CA: Plural Publishing, 2006), 1–12.

9. Tom Koch, *Thieves of Virtue: When Bioethics Stole Medicine* (Cambridge, MA: MIT Press, 2014).

10. Angeles Tan Alora and Josephine M. Lumitao, "An Introduction to an Authentically Non-Western Bioethics," in *Beyond a Western Bioethic: Voices from the Developing World*, ed. Angeles Tan Alora and Josephine M. Lumitao (Washington, DC: Georgetown University Press, 2001), 3–19.

11. Karen Peterson-Iyer, "Pharmacogenomics, Ethics, and Public Policy," *Kennedy Institute of Ethics Journal* 18, no. 1 (2008): 46.

12. See David Hollenbach, "The Common Good as Participation in Community: A Theological/Ethical Reflection on Some Empirical Issues," *Center for Advanced Catholic Studies*, University of Southern California (June 25–28, 2014): 15–18.

13. Pierce and Jameton, "Sustainable Health Care," 367.

14. Pierce and Jameton, "Sustainable Health Care," 367.

15. Jennifer Girod, "A Sustainable Medicine: Lessons from the Old Order Amish," *Journal of Medical Humanities* 23, no. 1 (2002): 34.

16. Anne Schwenkenbecher, "Is There an Obligation to Reduce One's Individual Carbon Footprint?," *Critical Review of International Social and Political Philosophy* 17, no. 2 (2014): 172.

17. World Health Organization, *Global Health Risks*.

18. Davidson R. Gwatkin, Abbas Bhuiya, and Cesar G. Victora, "Making Health Systems More Equitable," *Lancet* 364, no. 9441 (2004): 1275.

19. Pierce and Jameton, "Sustainable Health Care," 367.

20. Julien S. Murphy, "Is Pregnancy Necessary? Feminist Concerns about Ectogenesis," *Hypatia: Journal of Feminist Philosophy* 4, no. 3 (1989): 81.

21. This does not include non-reporting clinics. Centers for Disease Control and Prevention, American Society for Reproductive Medicine, Society for Assisted Reproductive Technology, *2015 Assisted Reproductive Technology National Summary Report* (Atlanta: US Department of Health and Human Services, 2017), 3.

22. Mary Anne Crandall, *Human Reproductive Technologies: Products and Global Markets*

(Wellesley, MA: BCC Research Market Forecasting, 2013).

23. Frances Price, "The Management of Uncertainty in Obstetric Practice: Ultrasonography, In Vitro Fertilisation, and Embryo Transfer," in *The New Reproductive Technologies*, ed. Maureen McNeil, Ian Varcoe, and Steven Yearley (London: Macmillan, 1990), 123–53.

24. Viv Groskop, "Do You Really Need a 'Mommy Makeover'?," *Guardian*, August 4, 2008.

25. Sylvia Burrow, "On the Cutting Edge: Ethical Responsiveness to Cesarean Rates," *American Journal of Bioethics* 12, no. 7 (2012): 44–52.

26. Karen N. Peart, "C-sections Linked to Breathing Problems in Preterm Infants," *Yale News*, February 10, 2010.

27. Luz Gibbons et al., *The Global Numbers and Costs of Additionally Needed and Unnecessary Caesarean Sections Performed per Year: Overuse as a Barrier to Universal Coverage: World Health Report* (Geneva: World Health Organization, 2010), 18.

28. Associated Press, "Same Sex Couples to Get IVF Help in Greater Manchester," *BBC News*, December 13, 2010. In France, the government will refuse fertility treatments to same-sex couples, single women, and the obese. Nicola Hebden, "Gay Couples 'Could Be Given Fertility Treatment,'" *The Local* (France), October 12, 2012.

29. "State Infertility Insurance Laws," American Society for Reproductive Medicine, https://www.reproductivefacts.org. Cristina Richie, "Reading Between the Lines: Infertility and Current Health Insurance Policies in the United States," *Clinical Ethics* 9, no. 4 (2014): 127–34.

30. Landon Trost and Robert Brannigan, "Oncofertility and the Male Cancer Patient," *Current Treatment Options in Oncology* 13, no. 2 (2012): 146–60.

31. There is a disconcerting connection between pornography use and fertility clinics. Associated Press, "NHS Criticised for Supplying Pornography to IVF Couples," *Independent U.K.*, September 8, 2010; Cristina Richie, "Feminist Bioethics, Pornography, and the Reproductive Technologies Business," *Blog of IJFAB: The International Journal of Feminist Approaches to Bioethics*, October 5, 2015.

32. Sumer Allensworth Wallace, Kiara L. Blough, and Laxmi A. Kondapalli, "Fertility Preservation in the Transgender Patient: Expanding Oncofertility Care beyond Cancer," *Gynecological Endocrinology* 30, no. 12 (2014): 868–71.

33. Dominic Stoop, Ana Cobo, and Sherman Silber, "Fertility Preservation for Age-Related Fertility Decline," *Lancet* 384, no. 9950 (2014): 1311–19.

34. Sénat de Belgique, "La loi du 6 juillet 2007 relative à la procréation médicalement

assistée et à la destination des embryons surnuméraires et des gametes," *Moniteur belge* 38575 (2007): art. 15.

35. Clarisa R. Gracia, Jorge J. E. Gracia, and Shasha Chen, "Ethical Dilemmas in Oncofertility: An Exploration of Three Clinical Scenarios," in *Oncofertility: Ethical, Legal, Social, and Medical Perspectives*, ed. Teresa Woodruff et al. (New York: Springer, 2010), 198.

36. Gwendolyn P. Quinn et al., "Preserving the Right to Future Children: An Ethical Case Analysis," *American Journal of Bioethics* 12, no. 6 (2012): 38–43.

37. Laurie Zoloth and Alyssa Henning, "Bioethics and Oncofertility: Arguments and Insights from Religious Traditions," in *Oncofertility: Ethical, Legal, Social, and Medical Perspectives*, ed. Teresa Woodruff et al. (New York: Springer, 2010), 264.

38. "In Vitro Fertilization: IVF," American Pregnancy Association, http://americanpregnancy.org.

39. Pascale Jadoul et al., "Efficacy of Ovarian Tissue Cryopreservation for Fertility Preservation: Lessons Learned from 545 Cases," *Human Reproduction* 32, no. 5 (2017): 1046–54.

40. Yechiel Mor and Joseph G. Schenker, "Ovarian Hyperstimulation Syndrome and Thrombotic Events," *American Journal of Reproductive Immunology* 72, no. 6 (2014): 541–48.

41. Mor and Schenker, "Ovarian Hyperstimulation Syndrome," 541–48.

42. M. Reigstad et al., "Cancer Risk among Parous Women Following Assisted Reproductive Technology," *Human Reproduction* 30, no. 8 (2015): 1952–63.

43. François Olivennes, "Avoiding Multiple Pregnancies in ART Double Trouble: Yes a Twin Pregnancy Is an Adverse Outcome," *Human Reproduction* 15, no. 8 (2000): 1661–63.

44. Natasa Tul et al., "The Contribution of Twins Conceived by Assisted Reproduction Technology to the Very Preterm Birth Rate: A Population-Based Study," *European Journal of Obstetrics & Gynecology and Reproductive Biology* 171, no. 2 (2013): 311–13.

45. Joyce Martin et al., "Births: Final Data for 2013," *National Vital Statistics Reports* 64, no. 1 (2015): 3.

46. Danielly S. Santana et al., "Twin Pregnancy and Severe Maternal Outcomes: The World Health Organization Multicountry Survey on Maternal and Newborn Health," *Obstetrics & Gynecology* 127, no. 4 (2016): 631–41.

47. Michèle Hansen et al., "Twins Born Following Assisted Reproductive Technology:

Perinatal Outcome and Admission to Hospital," *Human Reproduction* 24, no. 9 (2009): 2321–31.

48. Hansen et al., "Twins Born Following," 2330.

49. Amir Kugelman et al., "Iatrogenesis in Neonatal Intensive Care Units: Observational and Interventional, Prospective, Multicenter Study," *Pediatrics* 122, no. 3 (2008): 550–55. Gautham Suresh et al., "Voluntary Anonymous Reporting of Medical Errors for Neonatal Intensive Care," *Pediatrics* 113, no. 6 (2004): 1609–18.

50. Martha Nussbaum, "Aristotelian Social Democracy," in *Liberalism and the Good*, ed. R. Bruce Douglass, Gerald M. Mara, and Henry S. Richardson (New York: Routledge, 1990), 238.

51. World Health Organization, *Health Workforce* (Geneva: WHO Global Atlas of the Health Workforce, 2010).

52. World Health Organization, *Density of Doctors, Nurses and Midwives in the 49 Priority Countries* (Geneva: WHO Global Atlas of the Health Workforce, 2010).

53. "Health Service Coverage: Data by Country: Births Attended by Skilled Health Personnel," World Health Organization, http://www.who.int.

54. World Health Organization, *Global Health Risks*, 10.

55. Anne Donchin, "In Whose Interest? Policy and Politics in Assisted Reproduction," *Bioethics* 25, no. 2 (2011): 100.

56. Charlotte E. Warren et al., "'Sickness of Shame': Investigating Challenges and Resilience among Women Living with Obstetric Fistula in Kenya," in *Global Perspectives on Women's Sexual and Reproductive Health across the Lifecourse*, ed. Shonali Choudhury, Jennifer Toller Erausquin, and Mellissa Withers (New York: Springer, 2018), 91–109.

57. Marjorie Koblinsky et al., "Quality Maternity Care for Every Woman, Everywhere: A Call to Action," *Lancet* 388, no. 10057 (2016): 2307–20.

58. James Gallagher, "First Human Eggs Grown in Laboratory," *BBC News*, February 9, 2018.

59. Tim Jackson, "Live Better by Consuming Less: Is There a 'Double Dividend' in Sustainable Consumption?," *Journal of Industrial Ecology* 9, no. 1–2 (2005): 20.

60. Aparajita Dasgupta and Soumya Deb, "Telemedicine: A New Horizon in Public Health in India," *Indian Journal of Community Medicine* 33, no. 1 (2008): 3–8.

61. American Telemedicine Association, "What is Telemedicine?," http://www.americantelemed.org.

62. Randy S. Wax, "Canada: Where Are We Going?," in *ICU Resource Allocation in the New Millennium: Will We Say "No"?*, ed. David Crippen (New York: Springer, 2013), 126.

63. National Institute for Health Research, *The NIHR Carbon Reduction Guidelines* (England: NIHR, October 2010), 12.

64. National Health Service Sustainable Development Unit, *Saving Carbon, Improving Health: NHS Carbon Reduction Strategy for England* (London: NHS Sustainable Development Unit, 2009), 48.

65. Due to glitches in the telemedical systems that control patient records and medical prescriptions, several people have died. Christopher Rowland, "Hazards Tied to Medical Records Rush," *Boston Globe*, July 20, 2014.

66. "EMA HER," Modernizing Medicine, https://www.modmed.com.

67. Andrew Thorniley, "United Kingdom: Where Are We Going?," in *ICU Resource Allocation in the New Millennium: Will We Say "No"?*, ed. David Crippen (New York: Springer, 2013), 181; italics in the original.

68. Dasgupta and Deb, "Telemedicine," 3–8.

69. Dasgupta and Deb, "Telemedicine," 3–8.

70. Dasgupta and Deb, "Telemedicine," 3–8.

71. "About," World Health Partners, http://worldhealthpartners.org.

72. Health Research for Action, UC Berkeley, "Telemedicine Social Franchising in Rural Uttar Pradesh, India," 2012, http://www.healthresearchforaction.org.

73. Health Research for Action, "Telemedicine Social Franchising." Couple years of contraception, or couple-years of protection (CYP) refer to "the estimated protection provided by contraceptive methods during a one-year period." Jacqueline E. Darroch and Susheela Singh, *Estimating Unintended Pregnancies Averted by Couple-Years of Protection (CYP)* (New York: Guttmacher Institute, 2011).

74. Aetna Teledoc, https://www.teladoc.com/aetna. John Bohannon, "The Synthetic Therapist," *Science* (July 17, 2015): 250–51; Michal Wallace, "Eliza, Computer Therapist," Manifestation, http://www.manifestation.com. Eric Horvitz and Deirdre Mulligan, "Data, Privacy, and the Greater Good," *Science* (July 17, 2015): 253–55.

75. Dasgupta and Deb, "Telemedicine," 3–8.

76. José Goldemberg, "Leapfrog Energy Technologies," *Energy Policy* 26, no. 10 (1998): 729–41.

77. Nancy Kass, "Public Health Ethics: From Foundations and Frameworks to Justice and Global Public Health," *Journal of Law, Medicine and Ethics* 32, no. 2 (2004): 236.

CHAPTER 4. RESOURCE CONSERVATION

1. Jessica Pierce and Andrew Jameton, "Sustainable Health Care and Emerging Ethical Responsibilities," *Canadian Medical Association Journal* 164, no. 3 (2001): 367.

2. In economics, this is known as "efficiency." See Geoffrey Heal, "New Strategies for the Provision of Global Public Goods: Learning from International Environmental Challenges," in *Global Public Goods: International Cooperation in the 21st Century*, ed. Inge Kaul, Isabelle Grunberg, and Marc A. Stern (New York: United Nations Development Programme, 1999), 220–39.

3. Amartya Sen, "Equality of What?," Tanner Lectures on Human Values, Stanford University, May 22, 1979, 218.

4. Martha C. Nussbaum, "Human Functioning and Social Justice: In Defense of Aristotelian Essentialism," *Political Theory* 20, no. 2 (1992): 215.

5. Nussbaum, "Human Functioning and Social Justice," 216–21.

6. Nussbaum, "Human Functioning and Social Justice," 222.

7. Nussbaum, "Human Functioning and Social Justice," 214.

8. Joseph H. Howell and William Frederick Sale, "Specifying the Goals of Medicine," in *Life Choices: A Hastings Center Introduction to Bioethics*, 2nd ed., ed. Joseph H. Howell and William Frederick Sale (Washington, DC: Georgetown University Press, 2000), 62.

9. World Health Organization, *Preamble to the Constitution of the World Health Organization as Adopted by the International Health Conference, June 19–22, 1946*, Official Records of the World Health Organization, no. 2 (New York: World Health Organization, 1948), 100.

10. Howell and Sale, "Specifying the Goals of Medicine," 64.

11. Howell and Sale, "Specifying the Goals of Medicine," 67.

12. Howell and Sale, "Specifying the Goals of Medicine," 68–69.

13. Howell and Sale, "Specifying the Goals of Medicine," 71.

14. In less trivial situations, such as military deployment, menstrual suppression might be considered a health-care need that pertains to medical quality of life, especially when women are in extreme geographical locations without sanitation facilities. See Nicole Powell-Dunford et al., "Menstrual Suppression for Combat Operations: Advantages of Oral Contraceptive Pills," *Women's Health Issues* 21, no. 1 (2011): 86–91.

15. "Our Work: Medical Issues," Doctors Without Borders, https://www.doctorswithoutborders.org.

16. Marina Casini et al., "Why Teach 'Bioethics and Human Rights' to Healthcare Professions Undergraduates?," *JAHR: European Journal of Bioethics* 52, no. 10 (2014): 360.

17. Naomi Oreskes and Erik Conway, *Merchants of Doubt: How a Handful of Scientists Obscured the Truth on Issues from Tobacco Smoke to Global Warming* (London: Bloomsbury Publishing, 2010).

18. Andrea Vicini, "Is Transhumanism a Helpful Answer to Contemporary Bioethical Challenges?," lecture, Ethics Grand Rounds, University of Texas Southwestern Medical Center, Dallas, Texas, March 11, 2014, https://utswmed-ir.tdl.org/utswmed-ir/.

19. Erik Parens, "Is Better Always Good? The Enhancement Project," in *Enhancing Human Traits: Ethical and Social Implications*, ed. Erik Parens (Washington, DC: Georgetown University Press, 1998), 1–28.

20. Gerald P. McKenny, "Enhancements and the Quest for Perfection," *Christian Bioethics* 5, no. 2 (1999): 102.

21. Michael Hauskeller, "A Cure for Humanity: The Transhumanisation of Culture," *Trans-Humanities Journal* 8, no. 3 (2015): 131–47.

22. Nick Bostrom, "Human Genetic Enhancements: A Transhumanist Perspective," *Journal of Value Inquiry* 37, no. 4 (2003): 493–506.

23. Thomas Szasz, *The Medicalization of Everyday Life: Selected Essays* (Syracuse, NY: Syracuse University Press, 2007); Didier Fassin and Richard Rechtman, *The Empire of Trauma: An Inquiry into the Condition of Victimhood*, trans. Rachel Gomme (Princeton, NJ: Princeton University Press, 2009).

24. Howell and Sale, "Specifying the Goals of Medicine," 68–69.

25. Alexandre Baril and Kathryn Trevenen, "Exploring Ableism and Cisnormativity in the Conceptualization of Identity and Sexuality 'Disorders,'" *Annual Review of Critical Psychology* 11 (2014): 389–416; Eric Parens and Adrienne Asch, "The Disability Rights Critique of Prenatal Genetic Testing: Reflections and Recommendations," *Special Supplement Hastings Center Report* 29 (1999): S1–S22.

26. Daniel Callahan, "Sustainable Medicine," *Project Syndicate*, January 20, 2004.

27. Jean-Jacques Rousseau, *Discourse on the Origin of Inequality*, trans. Donald A. Cress (Indianapolis, IN: Hackett, 1992).

28. Collaborative Group on Epidemiological Studies of Ovarian Cancer, "Menopausal Hormone Use and Ovarian Cancer Risk: Individual Participant Meta-Analysis of 52

Epidemiological Studies," *Lancet* 385, no. 9980 (2015): 1835–42.

29. Henry Boardman et al., *Hormone Therapy for Preventing Cardiovascular Disease in Post-Menopausal Women* (Review) (New York: Wiley, 2015).

30. Pradeep K. Sacitharan, Sarah J. B. Snelling, and James R. Edwards, "Aging Mechanisms in Arthritic Disease," *Discovery Medicine* 14, no. 78 (2012): 345–52.

31. Sarah Derrett, Charlotte Paul, and Jenny M. Morris, "Waiting for Elective Surgery: Effects on Health-Related Quality of Life," *International Journal for Quality in Health Care* 11, no. 1 (1999): 47–57.

32. National Institute for Health and Care Excellence, "Offer Weight Loss Surgery to Obese People with Diabetes," *NICE*, November 27, 2014.

33. Jonathan Wilson et al., "Reducing the Risk of Major Elective Surgery: Randomised Controlled Trial of Preoperative Optimization of Oxygen Delivery," *BMJ* 318 (1999): 1099.

34. In 2015, a penis transplant raised ethical questions around offering elective, non-lifesaving surgeries that are medically risky. James Gallagher, "South Africans Perform First 'Successful' Penis Transplant," *BBC News*, March 13, 2015.

35. Paul Ehrlich, "Ecoethics: Now Central to All Ethics," *Bioethical Inquiry* 6 (2009): 417–36.

36. Paul Dolan et al., "QALY Maximisation and People's Preferences: A Methodological Review of the Literature," *Health Economics* 14, no. 2 (2005): 197–208.

37. Nisma Mujahid et al., "A UV-Independent Topical Small-Molecule Approach for Melanin Production in Human Skin," *Cell Reports* 19, no. 11 (2017): 2177–84.

38. "State Infertility Insurance Laws," American Society for Reproductive Medicine, https://www.reproductivefacts.org. Cristina Richie, "Reading Between the Lines: Infertility and Current Health Insurance Policies in the United States," *Clinical Ethics* 9, no. 4 (2014): 127–34.

39. Kristin Park, "Stigma Management among the Voluntarily Childless," *Sociological Perspectives* 45, no. 1 (2002): 21–45.

40. Julian Gill-Petersen, "The Value of the Future: The Child as Human Capital and the Neoliberal Labor of Race," *WSQ: Women's Studies Quarterly* 43, no. 1–2 (2015): 181–96.

41. Pierce and Jameton, "Sustainable Health Care," 366.

42. "Birth Control," United States Food and Drug Administration, https://www.fda.gov.

43. Burwell v. Hobby Lobby Stores, no. 13–354 (2014); Conestoga Wood Specialties Corp. v. Burwell, no. 13–356 (2014).

44. I. Glenn Cohen, Holly Fernandez Lynch, and Gregory D. Curfman, "When Religious Freedom Clashes with Access to Care," *New England Journal of Medicine* 371, no. 7 (2014): 598.

45. Joint United Nations Programme on HIV and AIDS, *Gap Report* (Geneva: UNAIDS, 2014), 234.

46. Cristina Richie, "Voluntary Sterilization for Childfree Women: Understanding Patient Profiles, Evaluating Accessibility, Examining Legislation," *Hastings Center Report* 43, no. 6 (2013): 36–44.

47. World Health Organization, "Maternal Mortality," http://www.who.int.

48. World Health Organization, "Maternal Mortality."

49. United Nations, *The Millennium Development Goals Report* (New York: United Nations, 2013), 33.

50. United Nations, *The Millennium Development Goals Report*, 28.

51. World Health Organization, "Maternal Mortality."

52. "The World FactBook: Country Comparison: Maternal Mortality Rates," Central Intelligence Agency, https://www.cia.gov.

53. Andreea A. Creanga et al., "Race, Ethnicity, and Nativity Differentials in Pregnancy-Related Mortality in the United States: 1993–2006," *Obstetrics & Gynecology* 120, no. 2, part 1 (2012): 261–68.

54. Mark Schiffman et al., "Human Papillomavirus and Cervical Cancer," *Lancet* 370, no. 9590 (2007): 890–907.

55. World Health Organization, *Global Health Risks: Mortality and Burden of Diseases Attributable to Selected Major Risks* (Geneva: WHO Press, 2009), 10.

56. Jane Dreaper, "The One Dollar Contraceptive Set to Make Family Planning Easier," *BBC News*, November 15, 2014.

57. David Crippen, *ICU Resource Allocation in the New Millennium: Will We Say "No"?* (New York: Springer, 2013), 338.

CHAPTER 5. SIMPLICITY

1. Arnold S. Relman, "The New Medical-Industrial Complex," *New England Journal of Medicine* 303, no. 17 (1980): 963–70.

2. Duane Elgin, *Voluntary Simplicity: Toward a Way of Life That Is Outwardly Simple, Inwardly Rich* (New York: Quill, 1993).

3. Emily Huddart Kennedy, Harvey Krahn, and Naomi Krogman, "Downshifting:

An Exploration of Motivation, Quality of Life, and Environmental Practices," *Sociological Forum* 28, no. 4 (2013): 764.

4. Tim Jackson, "Live Better by Consuming Less: Is There a 'Double Dividend' in Sustainable Consumption?," *Journal of Industrial Ecology* 9, no. 1–2 (2005): 23.

5. Francis O. Walker, "Essay Cultivating Simple Virtues in Medicine," Neurology 65, no. 10 (2005): 1678–80; Edmund Pellegrino, "Rationing Health Care: The Ethics of Medical Gatekeeping," *Journal of Contemporary Health Law and Policy* 2, no. 1 (1986): 23–46.

6. The Choosing Wisely campaign also highlights this approach. "Choosing Wisely," American Board of Internal Medicine Foundation, http://www.choosingwisely.org/.

7. Daniel Sulmasy and Beverly Moy, "Debating the Oncologist's Role in Defining the Value of Cancer Care: Our Duty Is to Our Patients," *Journal of Clinical Oncology* 32, no. 36 (2014): 4040.

8. See Edmund D. Pellegrino and David C. Thomasma, *A Philosophical Basis of Medical Practice: Toward a Philosophy and Ethic of the Healing Professions* (New York: Oxford University Press, 1981), 137, 139.

9. Sulmasy and Moy, "Debating the Oncologist's Role," 4040.

10. Sameer J. Patel et al., "Antibiotic Use in Neonatal Intensive Care Units and Adherence with Centers for Disease Control and Prevention 12 Step Campaign to Prevent Antimicrobial Resistance," *Pediatric Infectious Disease Journal* 28, no. 12 (2009): 1047.

11. Centers for Disease Control and Prevention, *Get Smart Health Care: Know When Antibiotics Work* (Atlanta: Centers for Disease Control and Prevention, n.d.).

12. National Quality Forum, National Quality Partners, and Antibiotic Stewardship Action Team, *National Quality Partners Playbook: Antibiotic Stewardship in Acute Care* (Washington, DC: National Quality Forum, 2016).

13. Gerald Dworkin, "Paternalism," in *Intervention and Reflection: Basic Issues in Medical Ethics*, 8th ed., ed. Ronald Munson (Australia: Thompson, 2008), 126, 128.

14. John Stuart Mill, *Utilitarianism and On Liberty*, ed. Mary Warnock (London: Fontana Press, 1962), 35.

15. "Infertility Definitions and Terminology," World Health Organization, http://www.who.int.

16. "Sexually Transmitted Diseases in the United States, 2008," Centers for Disease Control, https://www.cdc.gov.

17. Anne Donchin, "In Whose Interest? Policy and Politics in Assisted Reproduction," *Bioethics* 25, no. 2 (2011): 100.

18. Donchin, "In Whose Interest?," 100.

19. Donchin, "In Whose Interest?," 100.

20. Marie-Eve Lemoine and Vardit Ravitsky, "Sleepwalking into Infertility: The Need for a Public Health Approach towards Advanced Maternal Age," *American Journal of Bioethics* 15, no. 11 (2015): 37–48.

21. Cristina Richie, "What Would an Environmentally Sustainable Reproductive Technology Industry Look Like?," *Journal of Medical Ethics* 41, no. 5 (2015): 383–87.

22. Ann Oakley, *The Captured Womb: A History of the Medical Care of Pregnant Women* (Oxford: Basil Blackwell, 1984); Rebecca Kukla, *Mass Hysteria: Medicine, Culture, and Mothers' Bodies* (New York: Rowman and Littlefield, 2005); Adrienne Rich, *Of Woman Born: Motherhood as Institution and Experience* (New York: W.W. Norton, 1986). For opposing sides of the same debate, see Richard Johanson, Mary Newburn, and Alison Macfarlane, "Has the Medicalisation of Childbirth Gone Too Far?," *BMJ* 324, no. 7342 (2002): 892–95; Lachlan de Crespigny and Julian Savulescu, "Homebirth and the Future Child," *Journal of Medical Ethics* 40, no. 12 (2014): 807–12.

23. Barbara Andolsen, "Whose Sexuality? Whose Tradition? Women, Experience, and Roman Catholic Sexual Ethics," in *Readings in Moral Theology*, no. 9, *Feminist Ethics and the Catholic Moral Tradition*, ed. Charles Curran, Margaret Farley, and Richard McCormick (New York: Paulist Press, 1996), 225.

24. Eric Pianin and Brianna Ehley, "Budget Busting U.S. Obesity Costs Climb Past $300 Billion a Year," *Fiscal Times*, June 19, 2014.

25. Carl Lavie et al., "Body Composition and Heart Failure Prevalence and Prognosis: Getting to the Fat of the Matter in the 'Obesity Paradox,'" *Mayo Clinic Proceedings* 85, no. 7 (2010): 605–8.

26. Louise A. Baur, "Changing Perceptions of Obesity: Recollections of a Paediatrician," *Lancet* 378, no. 9793 (2011): 762–63.

27. F. Ofei, "Obesity—A Preventable Disease," *Ghana Medical Journal* 39, no. 3 (2005): 98.

28. Charles Tomson et al., "Race and End-Stage Renal Disease in the United States Medicare Population: The Disparity Persists," *Nephrology Carlton* 13, no. 7 (2008): 651–56. Emilie M. Townes, *Breaking the Fine Rain of Death: African American Health Issues and a Womanist Ethic of Care* (New York: Continuum, 1998).

29. Jean Pascal, Hélène Abbey-Huguenin, and Pierre Lombrail, "Inégalités sociales de santé: Quels impacts sur l'accès aux soins de prévention?," *Lien social et Politiques–RIAC 55, La santé au risque du social* (Spring 2006): 115–24.

30. *Lancet* 378, no. 9793 (2011): 743–847.

31. Steven L. Gortmaker et al., "Changing the Future of Obesity: Science, Policy, and Action," *Lancet* 378, no. 9793 (2011): 838–47.

32. Joseph Ax, "Bloomberg's Ban on Big Sodas Is Unconstitutional: Appeals Court," Reuters, July 30, 2013.

33. Cynthia L. Ogden et al., "Prevalence of Childhood and Adult Obesity in the United States, 2011–2012," *Journal of the American Medical Academy* 311, no. 8 (2014): 806–14.

34. BBC News: Health, "Child Obesity Rates 'On the Rise,'" *BBC News*, November 3, 2016.

35. Emma Rich, John Evans, and Laura De Pian, "Children's Bodies, Surveillance and the Obesity Crisis," in *Debating Obesity: Critical Perspectives*, ed. Emma Rich, Lee Monaghan, and Lucy Aphramor (New York: Palgrave Macmillan, 2010), 139–63.

36. Myriam Jacolin-Nackaerts and Jean Paul Clément, "La lutte contre l'obésité à l'école: Entre biopouvoir et individuation," *Lien social et Politiques*, no. 59 (2008): 47–60.

37. There are 3,500 calories in a pound. Daily caloric needs vary for individuals. The Mayo Clinic has a calorie calculator for those interested. "Calorie Calculator," Calculator.net, http://www.calculator.net.

38. Frédéric Picard and Leonard Guarente, "Calorie Restriction—the *SIR2* Connection," *Cell* 120, no. 4 (2005): 473–82.

39. Organisation Mondiale de la Santé, "Stratégie mondiale pour l'alimentation, l'exercice physique et la santé," Fifty-seventh World Health Assembly, A57/9 (17 April 2004): 1–24.

40. Daniel Callahan, "Obesity: Chasing an Elusive Epidemic," *Hastings Center Report* 43, no. 1 (2013): 34–40.

41. Garry Egger, "Personal Carbon Trading: A Potential 'Stealth Intervention' for Obesity Reduction?," *Medical Journal of Australia* 187, no. 3 (2007): 185–87.

42. F. Amianto et al., "The Forgotten Psychosocial Dimension of the Obesity Epidemic," *Lancet* 378, no. 9805 (2011): e8.

43. Kristin Voigt and Harald Schmidt, "Gastric Banding: Ethical Dilemmas in Reviewing Body Mass Index Thresholds," *Mayo Clinic Proceedings* 86, no. 10 (2011): 999.

44. Bruce Hensel, "Doctor Claims Tongue Patch Can Help Shed Pounds," *NBC Los Angeles*, April 6, 2011.

45. For this and other problems with "lap bands," see Voigt and Schmidt, "Gastric Banding," 999–1001.

46. U.S. Food and Drug Administration, "FDA Approves First-of-Kind Device to Treat Obesity," *ScienceDaily*, January 29, 2015.

47. L. J. Reece et al., "Use of Intragastric Balloons and a Lifestyle Support Programme to Promote Weight Loss in Severely Obese Adolescents: Pilot Study," *Appetite* 89 (2015): 305.

48. Committee on Approaching Death: Addressing Key End of Life Issues, *Dying in America: Improving Quality and Honoring Individual Preferences Near the End of Life* (Washington, DC: National Academies Press, 2015), ch. 2; Alfred F. Connors et al., "A Controlled Trial to Improve Care for Seriously Ill Hospitalized Patients: The Study to Understand Prognoses and Preferences for Outcomes and Risks of Treatments (SUPPORT)," *JAMA* 274, no. 20 (1995): 1591–98.

49. Carol Taylor and Robert Barnet, "Hand Feeding: Moral Obligation or Elective Intervention?," *Health Care Ethics USA* 22, no. 2 (2014): 14.

50. "1. A Clear and Specific Living Will Can Be Effective—If You Cannot Speak for Yourself," Caring Advocates, http://caringadvocates.org.

51. "Are These Goals Important to You?," Caring Advocates, http://caringadvocates.org.

52. Ian M. Seppelt, "Australia: Where Are We Going?," in *ICU Resource Allocation in the New Millennium: Will We Say "No"?*, ed. David Crippen (New York: Springer, 2013), 109.

53. Phil Perry, "State of the Art and Science Greener Clinics, Better Care," *Virtual Mentor: American Medical Association Journal of Ethics* 16, no. 9 (2014): 729.

54. "What Is Sustainable Medicine?," Kimberton Clinic, http://kimbertonclinic.com.

55. "About the Kimberton Clinic," Kimberton Clinic, http://kimbertonclinic.com.

56. Mary O'Brien, *The Politics of Reproduction* (Boulder, CO: Westview Press, 1989), 293.

57. Kennedy, Krahn, and Krogman, "Downshifting," 766.

58. Dorry de Beijer, "Motherhood and New Forms of Reproductive Technology: Passive Source of Nutrition and Rational Consumer," in *Motherhood: Experience, Institution, Theology*, ed. Anne Carr and Elizabeth Schussler Fiorenza (Edinburgh: T. and T. Clark, 1989), 75.

59. John Rawls, *Theory of Justice* (Cambridge, MA: Harvard University Press, 1971), 14.

CHAPTER 6. ETHICAL ECONOMICS

1. Guenter B. Risse, *Mending Bodies, Saving Souls: A History of Hospitals* (Oxford: Oxford University Press, 1999), 471–72.

2. Promises Austin offers a "60-Day Luxury Drug Rehabilitation Program"; "Treatment Programs," Promises Austin, https://www.promises.com. Walden Behavioral Care, LLC offers psychiatrists, dietitians, nurses, social workers, and mental health counselors as part of their services; "Adult Residential Eating Disorder Program," Walden Behavioral Care, https://www.waldeneatingdisorders.com. The Women's Wellness Institute of Dallas specializes in "vaginal rejuvenation," labiaplasty, and endometrial ablation "treatments" that scrape the uterus to reduce "heavy" menstruation; "Gynecology: Women's Health," Women's Wellness Institute of Dallas, https://www.womenswellnessinstitute.com.

3. Marcia Angell, *The Truth about Drug Companies: How They Deceive Us and What to Do About It* (New York: Random House, 2005), 161.

4. See Julia Whitty, "Diagnosing Health Care's Carbon Footprint," *Mother Jones*, November 10, 2009; United States Department of Commerce, International Trade Administration, *2016 Top Markets Report: Pharmaceuticals* (2016): 3.

5. Risse, *Mending Bodies, Saving Souls*, 474.

6. Elisabeth Rosenthal, "Is This a Hospital or a Hotel?," *New York Times*, September 21, 2013.

7. C. Ben Mitchell et al., *Biotechnology and the Human Good* (Washington, DC: Georgetown University Press, 2007), 123.

8. Amy Laura Hall, *Conceiving Parenthood: American Protestantism and the Spirit of Reproduction* (Grand Rapids, MI: Eerdmans, 2008), 147.

9. Cristina Richie, "Reading Between the Lines: Infertility and Current Health Insurance Policies in the United States," *Clinical Ethics* 9, no. 4 (2014): 127–34.

10. "Chance," ASB Bank New Zealand, November 14, 2010, https://www.youtube.com/watch?v=igcd4wNl3s8. For critical commentary, see Josephine Johnston, "Why I Mostly Love ASB Bank's IVF Ad," Hastings Center Bioethics Forum, December 20, 2010. https://www.thehastingscenter.org/why-i-mostly-love-asb-banks-ivf-ad/.

11. Lenworth M. Jacobs and Robert J. Schwartz, "The Impact of Prospective Reimbursement on Trauma Centers: An Alternative Payment Plan," *Archives of Surgery* 121, no. 4 (1986): 479–83. The independent business, Credit Bureau Systems, has established a financial payment strategy to "structure longer payment plans"

for hospital services rendered. Healthcare Affordable Repayment Plans, http://medicalrepayment.com.

12. Tim Jackson, "Live Better by Consuming Less: Is There a 'Double Dividend' in Sustainable Consumption?," *Journal of Industrial Ecology* 9, no. 1–2 (2005): 27.

13. "Biotechnology: Company Overview of Intercytex Limited," Bloomberg, https://www.bloomberg.com.

14. "Intercytex Phase II Hair Multiplication Trial Update," *Tressless News*, September 25, 2007.

15. "Intercytex Seeks £15m on AIM to Combat Baldness," *Business Weekly U.K.*, January 27, 2006.

16. David W. Crippen, "United States–Academic Medicine: Where Have We Been?," in *ICU Resource Allocation in the New Millennium: Will We Say "No"?*, ed. David Crippen (New York: Springer, 2013), 101.

17. Jackson, "Live Better," 24.

18. Michelle Roberts, "Viagra Can Be Sold Over the Counter," *BBC News*, November 28, 2017.

19. Wenjun Zhong et al., "Age and Sex Patterns of Drug Prescribing in a Defined American Population," *Mayo Clinic Proceedings* 88, no. 7 (2013): 697–707.

20. Jeanette W. Chung and David O. Meltzer, "Estimate of the Carbon Footprint of the US Health Care Sector," *Journal of the American Medical Association* 302, no. 18 (2009): 1971.

21. "Revenue of the Worldwide Pharmaceutical Market from 2001 to 2015 (in billion U.S. dollars)," Statista, https://www.statista.com.

22. Angell, *The Truth about Drug Companies*, 159.

23. United States Food and Drug Administration, *Approved Drug Products with Therapeutic Equivalence Evaluations*, 35th ed. (Silver Spring, MD: U.S. Food and Drug Administration, 2015).

24. Karen Peterson-Iyer, "Pharmacogenomics, Ethics, and Public Policy," *Kennedy Institute of Ethics Journal* 18, no. 1 (2008): 39.

25. Jane Wakefield, "First 3D-Printed Pill Approved by US Authorities," *BBC News*, August 4, 2015.

26. Henry I. Miller, "Bespoke Drugs," *Forbes*, January 9, 2009.

27. Felice J. Freyer, "Specialty Drugs Transform Lives—but at a Cost," *Boston Globe*, July 21, 2014.

28. Angell, *The Truth about Drug Companies.*

29. Hugh Pym, "Pfizer and Flynn Pharma Accused of Overcharging by CMA," *BBC News*, August 6, 2015.

30. See also Paul B. Ginsburg, *Wide Variation in Hospital and Physician Payment Rates Evidence of Provider Market Power* (Washington, DC: Center for Studying Health System Change, 2010).

31. Steven M. Jones et al., "Live Attenuated Recombinant Vaccine Protects Nonhuman Primates against Ebola and Marburg Viruses," *Nature Medicine* 11, no. 7 (2005): 786–90.

32. Denise Grady, "Ebola Vaccine, Ready for Test, Sat on the Shelf," *New York Times*, October 23, 2014.

33. "2014 Ebola Outbreak in West Africa—Case Counts," Centers for Disease Control, https://www.cdc.gov.

34. Christine Rushton, "Company Hikes Price 5,000% for Drug that Fights Complication of AIDS, Cancer," *USA Today*, September 18, 2015.

35. See also Wes Venteicher, "Insurer Lowers Cost of HIV Treatments after Discrimination Complaints," *Chicago Tribune*, March 27, 2015.

36. Colleen C. Denny and Ezekiel J. Emanuel, "US Health Aid beyond PEPFAR: The Mother and Child Campaign," *Journal of the American Medical Association* 300, no. 17 (2008): 2048–51.

37. Ernst Schumacher, *Small Is Beautiful: Economics as if People Mattered* (New York: Harper and Row, 1973); John B. Cobb, *Sustainability: Economics, Ecology and Justice* (Eugene, OR: Wipf and Stock Publishers, 2007); Paul Hawken, *The Ecology of Commerce: A Declaration of Sustainability* (New York: Harper Business, 2010); Tim Jackson, *Prosperity without Growth: Economics for a Finite Planet* (London: Earthscan, 2009).

38. Karl Marx, *Capital: An Abridged Edition*, ed. David McLellan (Oxford: Oxford World's Classics, 2008).

39. Kevin O'Rourke, *A Primer for Health Care Ethics: Essays for a Pluralistic Society*, 2nd ed. (Washington, DC: Georgetown University Press, 2000), 18.

40. Erin Lothes Biviano et al., "Is Fossil Fuel Investment a Sin?," *Health Care Ethics USA* 26, no. 1 (2018): 1–8.

41. David A. Crocker and Toby Linden, *Ethics of Consumption: The Good Life, Justice, and Global Stewardship* (Lanham, MD: Rowman & Littlefield, 1998).

42. David A. Crocker, "Consumption, Well-being, and Capability," in *Ethics of Consumption: The Good Life, Justice, and Global Stewardship*, ed. David A. Crocker and Toby Linden (Lanham, MD: Rowman & Littlefield, 1998), 366–90.

43. Robert Costanza et al., *An Introduction to Ecological Economics* (Boca Raton, FL: CRC Press, 2014); Herman E. Daly and Joshua Farley, *Ecological Economics: Principles and Applications* (Washington, DC: Island Press, 2011).

44. Frank Fraser Darling and John P. Milton, *Future Environments of North America* (Garden City, NY: Natural History Press, 1966), 717, quoted in Van Rensselaer Potter, *Bioethics: Bridge to the Future* (New Jersey: Prentice-Hall, 1971), 165.

45. Dan O'Neill, "Gross Domestic Product," in *Degrowth: A Vocabulary for a New Era*, ed. Giacomo D'Alisa, Federico Demaria, and Giorgos Kallis (New York: Routledge, 2014), 103.

46. Richard A. Easterlin, "Will Raising the Incomes of All Increase the Happiness of All?," *Journal of Economic Behavior & Organization* 27, no. 1 (1995): 35–47; Rafael Di Tella and Robert MacCulloch, "Gross National Happiness as an Answer to the Easterlin Paradox?," *Journal of Development Economics* 86, no. 1 (2008): 22–42.

47. Eric Davidson, *You Can't Eat GNP: Economics as if Ecology Mattered* (Cambridge, MA: Perseus Publishing, 2000).

48. Martha C. Nussbaum, "Human Functioning and Social Justice: In Defense of Aristotelian Essentialism," *Political Theory* 20, no. 2 (1992): 229.

49. Winton Bates, "Gross National Happiness," *Asian-Pacific Economic Literature* 23, no. 2 (2009): 1–16.

50. Nussbaum, "Human Functioning," 230.

51. Jenna Levy, "In U.S., Uninsured Rate Sinks to 12.9," *Gallup News*, January 7, 2015. The number was calculated using data from the United States Census Bureau. United States Census Bureau, "Census Bureau Projects U.S. and World Populations on New Year's Day," December 29, 2014, https://www.census.gov/newsroom.html.

52. United Nations Educational, Scientific and Cultural Organization, *Universal Declaration of Animal Rights* (Paris: UNESCO, 1978).

53. James A. Nash, "Biotic Rights and Human Ecological Responsibilities," in *The Annual of the Society of Christian Ethics*, ed. Harlan Beckley (Washington, DC: Georgetown University Press, 1993), 160.

54. Francisco De Vitoria, "On the American Indians," in *Vitoria Political Writings: Cambridge Texts in the History of Political Thought*, ed. Anthony Pagden and Jeremy

Lawrance (Cambridge: Cambridge University Press, 1991), 231–92.

55. United Nations, *Universal Declaration of Human Rights* (Geneva: United Nations, 1948).

56. Meghan Clark, *The Vision of Catholic Social Thought: The Virtue of Solidarity and the Praxis of Human Rights* (Minneapolis: Fortress, 2014), 2.

57. Nash, "Biotic Rights," 143.

58. Andrew Jameton, "Outline of the Ethical Implications of Earth's Limits for Health Care," *Journal of Medical Humanities* 23, no. 1 (2002): 55.

59. Theodore Dalrymple, "Is There a 'Right' to Health Care?," *Wall Street Journal*, July 28, 2009.

60. Khalid Malik, *Human Development Report 2014: Sustaining Human Progress: Reducing Vulnerabilities and Building Resilience* (New York: United Nations Development Programme, 2014).

61. United Nations Development Programme, *Human Development Report 2016: Human Development for Everyone* (New York: United Nations Development Programme, 2016).

62. Associated Press in San Francisco, "Surgeon General: I Have No Regrets about Calling Gun Violence Public Health Issue," *The Guardian*, August 16, 2015.

63. See also Joao Biehl, "Pharmaceutical Governance," in *Global Pharmaceuticals: Ethics, Markets, Practices*, ed. Adriana Petryna, Andrew Lakoff, and Arthur Kleinman (Durham, NC: Duke University Press, 2006), 206–39.

64. Abstinence, condoms, clean needles, and pre-exposure prophylaxis (PrEP) are options to prevent HIV. New research is being done on HIV vaccines, such as antibody mediated protection (AMP). Lindsay E. Young et al., "PrEP Chicago: A Randomized Controlled Peer Change Agent Intervention to Promote the Adoption of Pre-Exposure Prophylaxis for HIV Prevention among Young Black Men Who Have Sex with Men," *Clinical Trials* 15, no. 1 (2018): 44–52. Lynn Morris and Nonhlanhla N. Mkhize, "Prospects for Passive Immunity to Prevent HIV Infection," *PLoS Medicine* 14, no. 11 (2017): e1002436.

65. Celia W. Dugger, "Clinton Gets Five Companies to Reduce the Cost of AIDS Tests," *New York Times*, January 15, 2004.

66. Jane Galvão, "Brazil and Access to HIV/AIDS Drugs: A Question of Human Rights and Public Health," *American Journal of Public Health* 95, no. 7 (2005): 1111.

67. Galvão, "Brazil and Access to HIV/AIDS Drugs," 1111.

68. Galvão, "Brazil and Access to HIV/AIDS Drugs," 1110.

69. Rubens Costa-Filho, "Brazil: Where Are We Going?," in *ICU Resource Allocation in the New Millennium: Will We Say "No"?*, ed. David W. Crippen (New York: Springer, 2013), 119.

70. Galvão, "Brazil and Access to HIV/AIDS Drugs," 1113.

71. Galvão, "Brazil and Access to HIV/AIDS Drugs," 1113.

72. United Nations Programme on HIV and AIDS, http://www.unaids.org. See Jon Fuller and James Keenan, "Educating in a Time of HIV/AIDS: Learning from the Legacies of Human Rights, the Common Good, and the Works of Mercy," in *Opening Up: Speaking Out in the Church*, ed. Julian Filochowski and Peter Stanford (London: Darton, Longman & Todd, 2005), 95–113. See also Tracy Kidder, *Mountains beyond Mountains* (New York: Random House, 2009).

73. "About Us," Doctors Without Borders, https://www.doctorswithoutborders.org.

74. "Our Work: Medical Issues," Doctors Without Borders, https://www.doctorswithoutborders.org.

75. "Founding of MSF: People First," Doctors Without Borders, https://www.doctorswithoutborders.org.

76. Alyson Shontell, "If You Look Like This, Your Pay Check Will Be Higher Than Average," *Business Insider*, February 26, 2011.

77. Jessica Pierce and Andrew Jameton, "Sustainable Health Care and Emerging Ethical Responsibilities," *Canadian Medical Association Journal* 164, no. 3 (2001): 366.

78. Potter, *Bioethics*, 168.

79. Travis Bradberry writes, "Business is, after all, about making a profit." Travis Bradberry, "7 Ways to Blow Your Boss's Mind," *Forbes*, September 29, 2015.

CHAPTER 7. THE GREEN PATIENT

1. 350.org, https://350.org.

2. Arthur W. Perry, *Straight Talk about Cosmetic Surgery* (New Haven, CT: Yale University Press, 2007), xi.

3. Ty Kiisel, "You Are Judged by Your Appearance," *Forbes*, March 20, 2013.

4. Quoting the Association of Aesthetics 2002 Breast Surgery report, in C. Ben Mitchell et al., *Biotechnology and the Human Good* (Washington, DC: Georgetown University Press, 2007), 141.

5. Perry, *Straight Talk*, xi.

6. David W. Crippen, "United States–Academic Medicine: Where Have We Been?," in *ICU Resource Allocation in the New Millennium: Will We Say "No"?*, ed. David W. Crippen (New York: Springer, 2013), 101.

7. Jessica Pierce and Andrew Jameton, "Sustainable Health Care and Emerging Ethical Responsibilities," *Canadian Medical Association Journal* 164, no. 3 (2001): 376.

8. Marie Catherine Letendre and Joseph Tham, "Family and Healthcare Decision Making: Implications for Bioethics in China," *Studia Bioethica* 4, no. 3 (2011): 26–27.

9. Pierce and Jameton, "Sustainable Health Care," 376.

10. Jennifer Girod, "A Sustainable Medicine: Lessons from the Old Order Amish," *Journal of Medical Humanities* 23, no. 1 (2002): 31–42.

11. National Health Service Sustainable Development Unit, *Saving Carbon, Improving Health: NHS Carbon Reduction Strategy for England* (London: NHS Sustainable Development Unit, 2009), 8; emphasis mine.

12. Daniel P. Sulmasy, "Dignity and Bioethics: History, Theory, and Selected Applications," in *Human Dignity and Bioethics: Essays Commissioned by the President's Council on Bioethics*, ed. Adam Schulman (Washington, DC: President's Council on Bioethics, 2008): 469–501.

13. Aristotle, *Nicomachean Ethics*, trans. Martin Ostwald (New York: Macmillan, 1962), book 3, ch. 5, 114a.

14. Wenjun Zhong et al., "Age and Sex Patterns of Drug Prescribing in a Defined American Population," *Mayo Clinic Proceedings* 88, no. 7 (2013): 697–707.

15. Pierce and Jameton, "Sustainable Health Care," 365.

16. Thomas Szasz, *The Medicalization of Everyday Life: Selected Essays* (Syracuse, NY: Syracuse University Press, 2007), 167.

17. Michael Yeo, "Toward an Ethic of Empowerment for Health Promotion," *Health Promotion International* 8, no. 3 (1993): 225–35.

18. Terry L. Anderson and Donald R. Leal, *Free Market Environmentalism* (Boulder, CO: Westview Press, 1991).

19. "About Our Fees," Planned Parenthood of Central and Western New York, Inc., https://www.plannedparenthood.org/planned-parenthood-central-western-new-york/.

20. "Learn," Planned Parenthood, https://www.plannedparenthood.org.

21. Larry L. Rasmussen, "Next Journey: Sustainability for Six Billion and More," in *Ethics for a Small Planet: New Horizons on Population, Consumption, and Ecology*, ed. Daniel

C. Maguire and Larry L. Rasmussen (New York: State University of New York, 1998), 123.

CHAPTER 8. THE GREEN DOCTOR

1. Martin Donohoe and Gordon Schiff, "A Call to Service: Social Justice Is a Public Health Issue," *Virtual Mentor: American Medical Association Journal of Ethics* 16, no. 9 (2014): 702.

2. "Density of Physicians (total number per 1000 population)," World Health Organization, http://www.who.int.

3. Suwit Wibulpolprasert and Paichit Pengpaibon, "Integrated Strategies to Tackle the Inequitable Distribution of Doctors in Thailand: Four Decades of Experience," *Human Resources for Health* 1, no. 1 (2003): 12.

4. Onyebuchi A. Arah, Uzor C. Ogbu, and Chukwudi E. Okeke, "Too Poor to Leave, Too Rich to Stay: Developmental and Global Health Correlates of Physician Migration to the United States, Canada, Australia, and the United Kingdom," *American Journal of Public Health* 98, no. 1 (2008): 148.

5. Nir Eyal and Till Bärnighausen, "Precommitting to Serve the Underserved," *American Journal of Bioethics* 12, no. 5 (2012): 23–34.

6. Onyebuchi Arah, "The Metrics and Correlates of Physician Migration from Africa," *BMC Public Health* 7, no. 1 (2007): 83.

7. Karen Peterson-Iyer, "Pharmacogenomics, Ethics, and Public Policy," *Kennedy Institute of Ethics Journal* 18, no. 1 (2008): 39.

8. Matthew Wynia, "Advocate as a Doctor or Advocate as a Citizen?," *Virtual Mentor: American Medical Association Journal of Ethics* 16, no. 9 (2014): 694–98.

9. Jonathan Wilson et al., "Reducing the Risk of Major Elective Surgery: Randomised Controlled Trial of Preoperative Optimization of Oxygen Delivery," *BMJ* 318, no. 7191 (1999): 1100.

10. Thomas Kerz, "Germany: Where Are We Going?," in *ICU Resource Allocation in the New Millennium: Will We Say "No"?*, ed. David W. Crippen (New York: Springer, 2013), 135.

11. Kerz, "Germany," 135.

12. R. Eric Hodgson and Timothy C. Hardcastle, "South Africa: Where Have We Been?," in *ICU Resource Allocation in the New Millennium: Will We Say "No"?*, ed. David W. Crippen (New York: Springer, 2013), 80.

13. Annette Mendola et al., "Worse Than Futile: Medically Non-Beneficial Treatment in

the Setting of Complicated Grief," 11th Annual International Conference on Clinical Ethics and Consultation, New York City, May 21, 2015.

14. American Psychiatric Association, *Diagnostic and Statistical Manual of Mental Disorders*, 5th ed. (Washington, DC: American Psychiatric Association, 2013).

15. Transgender Law Center, *10 Tips for Working with Transgender Patients*, http://transgenderlawcenter.org.

16. Alexandre Baril, "Needing to Acquire a Physical Impairment/Disability: (Re)Thinking the Connections between Trans and Disability Studies through Transability," trans. Catriona LeBlanc, *Hypatia: Journal of Feminist Philosophy* 30, no. 1 (2015): 30–48; Jacqueline K. Hewitt et al., "Hormone Treatment of Gender Identity Disorder in a Cohort of Children and Adolescents," *Medical Journal of Australia* 196, no. 9 (2012): 578–81.

17. Miroslav Djordjevic et al., "Reversal Surgery in Regretful Male-to-Female Transsexuals after Sex Reassignment Surgery," *Journal of Sexual Medicine* 13, no. 6 (2016): 1000–1007; Gunter Heylens et al., "Transgender Persons Applying for Euthanasia in Belgium: A Case Report and Implications for Assessment and Treatment," *Journal of Psychiatry* 18, no. 6 (2015): 1000347.

18. Two landmark cases in 2014 demonstrated the growing acceptance of, and vocal political lobby for, sex reassignment surgery for those on Medicaid and for violent prisoners. See Department of Health and Human Services, Departmental Appeals Board Appellate Division NCD 140.3, Transsexual Surgery Docket no. A-13–87, Decision No. 2576, May 30, 2014; Ryan Parker, "Federal Judge Orders California Prison Inmate Be Granted Sexual Reassignment," *Los Angeles Times*, April 2, 2015.

19. Cristina Richie, "A Queer, Feminist Bioethics Critique of Facial Feminization Surgery," *American Journal of Bioethics* 18, no. 12 (2018): 33–35.

20. Judith Halberstam, "F2M: The Making of Female Masculinity," in *Feminist Theory and the Body: A Reader*, ed. Janet Price and Margrit Shildrick (New York: Routledge, 1999), 125–33.

21. Kristin Voigt and Harald Schmidt, "Gastric Banding: Ethical Dilemmas in Reviewing Body Mass Index Thresholds," *Mayo Clinic Proceedings* 86, no. 10 (2011): 1000.

22. Voigt and Schmidt, "Gastric Banding," 1000.

23. Anne Lawrence, "Factors Associated with Satisfaction or Regret Following Male-to-Female Sex Reassignment Surgery," *Archives of Sexual Behavior* 32, no. 4 (2003): 299–315.

24. See Laura Purdy, "A Bioethics Perspective on Sex Reassignment Therapy," unpublished manuscript, April 2015, 4–7, https://www.academia.edu/11995270.

25. Tracy Miller, "Belgian Transsexual Dies by Euthanasia after Unsatisfactory Sex Change Operation," *New York Daily News*, October 2, 2013.

26. Annelou L. C. De Vries et al., "Young Adult Psychological Outcome after Puberty Suppression and Gender Reassignment," *Pediatrics* 134, no. 4 (2014): 696–704.

27. Rubens Costa-Filho, "Brazil: Where Are We Going?," in *ICU Resource Allocation in the New Millennium: Will We Say "No"?*, ed. David W. Crippen (New York: Springer, 2013), 117.

28. Costa-Filho, "Brazil," 116.

29. "Dental Implants Facts and Figures," American Academy of Implant Dentistry, https://www.aaid.com/index.html.

30. Centers for Disease Control and Prevention, American Society for Reproductive Medicine, Society for Assisted Reproductive Technology, *2015 Assisted Reproductive Technology National Summary Report* (Atlanta: US Department of Health and Human Services, 2017), 3.

31. "What Is Assisted Reproductive Technology? Most Recent ART Data," Centers for Disease Control, https://www.cdc.gov.

32. "Statistics," American Society for Aesthetic Plastic Surgery, https://www.surgery.org.

33. Costa-Filho, "Brazil," 104.

34. Jessica Pierce and Andrew Jameton, "Sustainable Health Care and Emerging Ethical Responsibilities," *Canadian Medical Association Journal* 164, no. 3 (2001): 366.

CHAPTER 9. THE GREEN HEALTH-CARE PLAN

1. Ronald Bayer and Amy L. Fairchild, "The Genesis of Public Health Ethics," *Bioethics* 18, no. 6 (2004): 489; Lawrence O. Gostin, "Public Health Law in an Age of Terrorism: Rethinking Individual Rights and Common Goods," *Health Affairs* 21, no. 6 (2002): 79–93.

2. Thomas Kerz, "Germany: Where Are We Going?," in *ICU Resource Allocation in the New Millennium: Will We Say "No"?*, ed. David W. Crippen (New York: Springer, 2013), 133.

3. Kerz, "Germany," 133.

4. Oregon Health Plan, "Prioritized List of Health Services," January 1, 2015, http://www.oregon.gov.

5. Leonard M. Fleck, *Just Caring: Health Care Rationing and Democratic Deliberation* (New York: Oxford University Press, 2009); Rebecca A. Bruni, Andreas Laupacis, and Douglas K. Martin, "Public Engagement in Setting Priorities in Health Care," *Canadian Medical Association Journal* 179, no. 1 (2008): 15–18; Hillary Wicai Viers, "What Is Democratic Deliberation? A Q&A with Bioethics Commission Chair Amy Gutmann," *Blog of the Presidential Commission for the Study of Bioethical Issues,* September 10, 2014, https://bioethicsarchive.georgetown.edu.

6. National Health Service Sustainable Development Unit, *Saving Carbon, Improving Health: NHS Carbon Reduction Strategy for England* (London: NHS Sustainable Development Unit, 2009), 15.

7. Geoff Herbert, "Buffalo Teachers Still Get Free Plastic Surgery Courtesy of Taxpayers," Syracuse.com, January 19, 2012. "New Mothers: Get Your Breast Pump Cost-Free, Delivered Right to Your Door," Blue Cross Blue Shield of Massachusetts, https://myblue.bluecrossma.com/healthy-living/.

8. Toni Clarke and Ransdell Pierson, "FDA Approves 'Female Viagra' with Strong Warning," Reuters, August 19, 2015.

9. David Crippen, "United States–Academic Medicine: Where Have We Been?," in *ICU Resource Allocation in the New Millennium: Will We Say "No"?*, ed. David W. Crippen (New York: Springer, 2013), 105.

10. Anthony Costello et al., "Managing the Health Effects of Climate Change," *Lancet* 373, no. 9676 (2009): 1708.

11. Stephen Streat, "New Zealand: Where Have We Been?," in *ICU Resource Allocation in the New Millennium: Will We Say "No"?*, ed. David W. Crippen (New York: Springer, 2013), 70.

12. See also Zosia Kmietowicz, "Trust Defers Surgical Referrals for Patients to Lose Weight and Stop Smoking," *BMJ: British Medical Journal* (*Online*) 343 (2011); Laura Donnelly, "NHS Provokes Fury with Indefinite Surgery Ban for Smokers and Obese," *Telegraph*, October 17, 2017.

13. Ian M. Seppelt, "Australia: Where Have We Been?," in *ICU Resource Allocation in the New Millennium: Will We Say "No"?*, ed. David W. Crippen (New York: Springer, 2013), 4.

14. Kerz, "Germany," 137.

15. Kerz, "Germany," 137.

16. Tom Beauchamp and James Childress, *Principles of Biomedical Ethics*, 4th ed. (New

York: Oxford University Press, 1994), 16.

17. Based on a 35-year old single female, with a $25,000 income in Pitt County, North Carolina. "Summary of Benefits and Coverage: Coverage for: Individual + Family Plan Type: PPO," Blue Cross Blue Shield of North Carolina, Blue Advantage Bronze 6650, Coverage Period: 01/01/2018–12/31/2018, available at https://www.bluecrossnc.com.

18. Inge Kaul, Isabelle Grunberg, and Mark Stern, eds., *Global Public Goods International Cooperation in the 21st Century* (Oxford: Oxford University Press, 1999).

19. Y. Claire Wang et al., "Health and Economic Burden of the Projected Obesity Trends in the USA and the UK," *Lancet* 378, no. 9793 (2011): 815–25.

20. Cynthia Ogden et al., *Prevalence of Obesity among Adults and Youth: United States, 2011–2014* (Hyattsville, MD: US Department of Health and Human Services, Centers for Disease Control and Prevention, National Center for Health Statistics, 2015).

21. Health Care Without Harm, *Healthier Hospitals Initiative* (Reston, VA: Health Care Without Harm, n.d.), 5.

22. American Sleep Apnea Association, *Choosing a PAP Machine* (Washington, DC: American Sleep Apnea Association, 2017).

23. Tat Chan, Chakravarthi Narasimhan, and Ying Xie, "Treatment Effectiveness and Side Effects: A Model of Physician Learning," *Management Science* 59, no. 6 (2013): 1309–25.

24. Alison Fildes et al., "Probability of an Obese Person Attaining Normal Body Weight: Cohort Study Using Electronic Health Records," *American Journal of Public Health* 105, no. 9 (2015): e54–59.

25. National Institute for Health and Care Excellence, "Offer Weight Loss Surgery to Obese People with Diabetes," *NICE*, November 27, 2014.

26. Kim Knowlton et al., "Six Climate Change–Related Events in the United States Accounted for about $14 Billion in Lost Lives and Health Costs," *Health Affairs* 30, no. 11 (2011): 2167–76.

27. Carl J. Lavie et al., "Obesity and Cardiovascular Diseases: Implications Regarding Fitness, Fatness, and Severity in the Obesity Paradox," *Journal of the American College of Cardiology* 63, no. 14 (2014): 1345–54.

28. Ian Roberts, "The NHS Carbon Reduction Strategy," *BMJ* 38, no. 7689 (2009): 248.

29. Daniel Callahan, "Obesity: Chasing an Elusive Epidemic," *Hastings Center Report* 43, no. 1 (2013): 34–40.

30. Yves Jalbert and Lyne Mongeau, "Prévenir l'obésité: Un aperçu des programmes, plans d'action, stratégies et politiques sur l'alimentation et la nutrition," *Institut national de santé publique du Québec* (2006): 1–28.

31. Henry Shue, "Subsistence Emissions and Luxury Emissions," *Law & Policy* 15, no. 1 (1993): 39–60.

32. "What Is Erectile Dysfunction?," Viagra, https://www.viagra.com.

33. Andy Miah and Emma Rich, *The Medicalization of Cyberspace* (New York: Routledge, 2008), 88. See also Leonore Tiefer, "The Viagra Phenomena," *Sexualities* 9, no. 3 (2006): 273–94.

34. Associated Press, "Viagra Is a $50 Million Pentagon Budget Item," *New York Times*, October 4, 1998.

35. Patricia Kime, "DoD Spends $84M a Year on Viagra, Similar Meds," *Military Times*, February 13, 2015.

36. Thomas Szasz, *The Medicalization of Everyday Life: Selected Essays* (Syracuse, NY: Syracuse University Press, 2007), xiv.

37. "Perks," Blue Cross Blue Shield of Massachusetts, http://www.studentbluema.com.

38. National Health Service Sustainable Development Unit, *Saving Carbon*, 45.

CONCLUSION

1. Louise P. King and Janet Brown, "Clinical Case: Educating Patients as Medicine Goes Green," *Virtual Mentor: American Medical Association Journal of Ethics* 11, no. 6 (2009): 427–33.

2. Cristina Richie, "What Would an Environmentally Sustainable Reproductive Technology Industry Look Like?," *Journal of Medical Ethics* 41, no. 5 (2015): 383–87; Laura Donnelly, "Single Women Should Not Get Free IVF, say Ethics Experts," *Telegraph*, July 25, 2014.

3. Jason Lee Fishel, "The Green Staff of Asclepius: Envisioning Sustainable Medicine," (PhD diss., University of Tennessee, Knoxville, 2014), 112.

4. World Health Organization, *Global Health Risks: Mortality and Burden of Diseases Attributable to Selected Major Risks* (Geneva: WHO Press, 2009), 24; Cheryl Cox Macpherson, "Climate Change Is a Bioethics Problem," *Bioethics* 27, no. 6 (2013): 305–8; Cheryl C. Macpherson and Muge Akpinar-Elci, "Caribbean Heat Threatens Health, Well-being and the Future of Humanity," *Public Health Ethics* 8, no. 2 (2015): 196–208.

5. Nancy Kass, "Public Health Ethics: From Foundations and Frameworks to Justice and Global Public Health," *Journal of Law, Medicine and Ethics* 32, no. 2 (2004): 235.

6. Macpherson and Akpinar-Elci, "Caribbean Heat Threatens," 204.

References

Alora, Angeles Tan, and Josephine M. Lumitao. "An Introduction to an Authentically Non-Western Bioethics." In *Beyond a Western Bioethic: Voices from the Developing World*, edited by Angeles Tan Alora and Josephine M. Lumitao, 3–19. Washington DC: Georgetown University Press, 2001.

American Psychiatric Association. *Diagnostic and Statistical Manual of Mental Disorders*. 5th ed. Washington, DC: American Psychiatric Association, 2013.

American Sleep Apnea Association. *Choosing a PAP Machine*. Washington, DC: American Sleep Apnea Association, 2017.

American Society for Reproductive Medicine. "State Infertility Insurance Laws." https://www.reproductivefacts.org.

Amianto, F., L. Lavagnino, G. Abbate-Daga, and S. Fassino. "The Forgotten Psychosocial Dimension of the Obesity Epidemic." *Lancet* 378, no. 9805 (2011): e8.

Anderson, Terry L., and Donald R. Leal. *Free Market Environmentalism*. Boulder, CO: Westview Press, 1991.

Andolsen, Barbara. "Whose Sexuality? Whose Tradition? Women, Experience, and Roman Catholic Sexual Ethics." In *Readings in Moral Theology*, no. 9, *Feminist Ethics and the Catholic Moral Tradition*, edited by Charles Curran, Margaret Farley, and Richard

McCormick, 207–39. New York: Paulist Press, 1996.

Angell, Marcia. *The Truth about Drug Companies: How They Deceive Us and What to Do About It.* New York: Random House, 2005.

Arah, Onyebuchi. "The Metrics and Correlates of Physician Migration from Africa." *BMC Public Health* 7, no. 1 (2007): 83.

Arah, Onyebuchi A., Uzor C. Ogbu, and Chukwudi E. Okeke. "Too Poor to Leave, Too Rich to Stay: Developmental and Global Health Correlates of Physician Migration to the United States, Canada, Australia, and the United Kingdom." *American Journal of Public Health* 98, no. 1 (2008): 148–54.

Aristotle. *Nicomachean Ethics.* Translated by Martin Ostwald. New York: Macmillan, 1962.

Aristotle. *Politics.* Translated by Ernest Barker. Oxford: Oxford University Press, 1995.

Associated Press. "NHS Criticised for Supplying Pornography to IVF Couples." *Independent U.K.,* September 8, 2010.

Associated Press. "Same Sex Couples to Get IVF Help in Greater Manchester." *BBC News,* December 13, 2010.

Associated Press. "Viagra is a $50 Million Pentagon Budget Item." *New York Times,* October 4, 1998.

Associated Press in San Francisco. "Surgeon General: I Have No Regrets about Calling Gun Violence Public Health Issue." *Guardian,* August 16, 2015.

Ax, Joseph. "Bloomberg's Ban on Big Sodas Is Unconstitutional: Appeals Court." Reuters, July 30, 2013.

Baril, Alexandre. "Needing to Acquire a Physical Impairment/Disability: (Re)Thinking the Connections between Trans and Disability Studies through Transability." Translated by Catriona LeBlanc. *Hypatia: Journal of Feminist Philosophy* 30, no. 1 (2015): 30–48.

Baril, Alexandre, and Kathryn Trevenen. "Exploring Ableism and Cisnormativity in the Conceptualization of Identity and Sexuality 'Disorders.'" *Annual Review of Critical Psychology* 11 (2014): 389–416.

Bates, Winton. "Gross National Happiness." *Asian-Pacific Economic Literature* 23, no. 2 (2009): 1–16.

Baur, Louise A. "Changing Perceptions of Obesity: Recollections of a Paediatrician." *Lancet* 378, no. 9793 (2011): 762–63.

Bayer, Ronald, and Amy L. Fairchild. "The Genesis of Public Health Ethics." *Bioethics* 18, no. 6 (2004): 473–92.

BBC News: Health. "Child Obesity Rates 'On the Rise.'" *BBC News,* November 3, 2016.

Beauchamp, Tom, and James Childress. *Principles of Biomedical Ethics*. 1st ed. New York: Oxford University Press, 1979.

Beauchamp, Tom, and James Childress. *Principles of Biomedical Ethics*. 4th ed. New York: Oxford University Press, 1994.

Beauchamp, Tom, and James Childress. *Principles of Biomedical Ethics*. 5th ed. New York: Oxford University Press, 2001.

Berners-Lee, Mike. *How Bad Are Bananas? The Carbon Footprint of Everything*. London: Profile Book, 2010.

Biehl, Joao. "Pharmaceutical Governance." In *Global Pharmaceuticals: Ethics, Markets, Practices*, edited by Adriana Petryna, Andrew Lakoff, and Arthur Kleinman, 206–39. Durham, NC: Duke University Press, 2006.

Biviano, Erin Lothes, Daniel DiLeo, Cristina Richie, and Tobias Winright. "Is Fossil Fuel Investment a Sin?" *Health Care Ethics USA* 26, no. 1 (2018): 1–8.

Boardman, Henry, Louise Hartley, Anne Eisinga, Caroline Main, Marta Roqué i Figuls, Xavier Bonfill Cosp, Rafael Gabriel Sanchez, and Beatrice Knight. *Hormone Therapy for Preventing Cardiovascular Disease in Post-Menopausal Women* (Review). New York: Wiley, 2015.

Boff, Leonardo. *Cry of the Earth, Cry of the Poor*. Maryknoll, NY: Orbis Books, 1997.

Bohannon, John. "The Synthetic Therapist." *Science* (July 17, 2015): 250–51.

Bostrom, Nick. "Human Genetic Enhancements: A Transhumanist Perspective." *Journal of Value Inquiry* 37, no. 4 (2003): 493–506.

Bradberry, Travis. "7 Ways To Blow Your Boss's Mind." *Forbes*, September 29, 2015.

Bradshaw, Corey J. A., and Barry W. Brook. "Human Population Reduction Is Not a Quick Fix for Environmental Problems." *PNAS: Proceedings of the National Academy of Sciences of the United States of America* (October 27, 2014): 16610–15.

Brundtland, Gro Harlem. *Report of the World Commission on Environment and Development: "Our Common Future."* New York: United Nations, 1987.

Bruni, Rebecca A., Andreas Laupacis, and Douglas K. Martin. "Public Engagement in Setting Priorities in Health Care." *Canadian Medical Association Journal* 179, no. 1 (2008): 15–18.

Burnett, Jill, M. Clarke, J. Darbyshire, A. Haines, I. Roberts, H. Shakur, N. Siegfried, and P. Wilkinson. "Towards Sustainable Clinical Trials." *BMJ* 334 (2007): 671–73.

Burrow, Sylvia. "On the Cutting Edge: Ethical Responsiveness to Cesarean Rates." *American Journal of Bioethics* 12, no. 7 (2012): 44–52.

Burwell v. Hobby Lobby Stores. No. 13–354 (2014).

Callahan, Daniel. "Obesity: Chasing an Elusive Epidemic." *Hastings Center Report* 43, no. 1 (2013): 34–40.

Callahan, Daniel. "Principlism and Communitarianism." *Journal of Medical Ethics* 29, no. 5 (2003): 287–91.

Callahan, Daniel. "Sustainable Medicine." *Project Syndicate*, January 20, 2004.

Carson, Rachel. *Silent Spring*. Boston: Houghton Mifflin, 1962.

Casini, Marina, Joseph Meaney, Emanuela Midolo, Anto Cartolovni, Dario Sacchini, and Antonio G. Spagnolo. "Why Teach 'Bioethics and Human Rights' to Healthcare Professions Undergraduates?" *JAHR: European Journal of Bioethics* 52, no. 10 (2014): 349–68.

Centers for Disease Control and Prevention. *Get Smart Health Care: Know When Antibiotics Work*. Atlanta: Centers for Disease Control and Prevention, n.d.

Centers for Disease Control and Prevention, American Society for Reproductive Medicine, Society for Assisted Reproductive Technology. *2015 Assisted Reproductive Technology National Summary Report*. Atlanta: US Department of Health and Human Services, 2017.

Chan, Tat, Chakravarthi Narasimhan, and Ying Xie. "Treatment Effectiveness and Side Effects: A Model of Physician Learning." *Management Science* 59, no. 6 (2013): 1309–25.

Chertow, Marian R. "The IPAT Equation and Its Variants." *Journal of Industrial Ecology* 4, no. 4 (2000): 13–29.

Chung, Jeanette W., and David O. Meltzer. "Estimate of the Carbon Footprint of the US Health Care Sector." *Journal of the American Medical Association* 302, no. 18 (2009): 1970–72.

Clark, Meghan. *The Vision of Catholic Social Thought: The Virtue of Solidarity and the Praxis of Human Rights*. Minneapolis: Fortress, 2014.

Clarke, Toni, and Ransdell Pierson. "FDA Approves 'Female Viagra' with Strong Warning." Reuters, August 19, 2015.

Cobb, John B. *Sustainability: Economics, Ecology and Justice*. Eugene, OR: Wipf and Stock Publishers, 2007.

Cohen, I. Glenn, Holly Fernandez Lynch, and Gregory D. Curfman. "When Religious Freedom Clashes with Access to Care." *New England Journal of Medicine* 371, no. 7 (2014): 596–99.

Collaborative Group on Epidemiological Studies of Ovarian Cancer. "Menopausal Hormone Use and Ovarian Cancer Risk: Individual Participant Meta-Analysis of 52 Epidemiological Studies." *Lancet* 385, no. 9980 (2015): 1835–42.

Committee on Approaching Death: Addressing Key End of Life Issues. *Dying in America: Improving Quality and Honoring Individual Preferences Near the End of Life.* Washington DC: National Academies Press, 2015.

Conestoga Wood Specialties Corp. v. Burwell. No. 13–356 (2014).

Connors, Alfred F., Neal V. Dawson, Norman A. Desbiens, William J. Fulkerson, Lee Goldman, William A. Knaus, Joanne Lynn et al. "A Controlled Trial to Improve Care for Seriously Ill Hospitalized Patients: The Study to Understand Prognoses and Preferences for Outcomes and Risks of Treatments (SUPPORT)." *JAMA* 274, no. 20 (1995): 1591–98.

Costa-Filho, Rubens. "Brazil: Where Are We Going?" In *ICU Resource Allocation in the New Millennium: Will We Say "No"?*, edited by David W. Crippen, 113–21. New York: Springer, 2013.

Costanza, Robert, John H. Cumberland, Herman Daly, Robert Goodland, Richard B. Norgaard, Ida Kubiszewski, and Carol Franco. *An Introduction to Ecological Economics.* Boca Raton, FL: CRC Press, 2014.

Costello, Anthony, Mustafa Abbas, Adriana Allen, Sarah Ball, Sarah Bell, Richard Bellamy, Sharon Friel et al. "Managing the Health Effects of Climate Change." *Lancet* 373, no. 9676 (2009): 1693–733.

Crandall, Mary Anne. *Human Reproductive Technologies: Products and Global Markets.* Wellesley, MA: BCC Research Market Forecasting, 2013.

Creanga, Andreea A., Cynthia J. Berg, Carla Syverson, Kristi Seed, F. Carol Bruce, and William M. Callaghan. "Race, Ethnicity, and Nativity Differentials in Pregnancy-Related Mortality in the United States: 1993–2006." *Obstetrics & Gynecology* 120, no. 2, part 1 (2012): 261–68.

Crippen, David. *ICU Resource Allocation in the New Millennium: Will We Say "No"?* New York: Springer, 2013.

Crippen, David W. "United States–Academic Medicine: Where Have We Been?" In *ICU Resource Allocation in the New Millennium: Will We Say "No"?*, edited by David Crippen, 101–15. New York: Springer, 2013.

Crocker, David A. "Consumption, Well-being, and Capability." In *Ethics of Consumption: The Good Life, Justice, and Global Stewardship*, edited by David A. Crocker and Toby

Linden, 366–90. Lanham, MD: Rowman & Littlefield, 1998.

Crocker, David A., and Toby Linden. *Ethics of Consumption: The Good Life, Justice, and Global Stewardship*. Lanham, MD: Rowman & Littlefield, 1998.

Dalrymple, Theodore. "Is There a 'Right' to Health Care?" *Wall Street Journal*, July 28, 2009.

Daly, Herman E., and Joshua Farley. *Ecological Economics: Principles and Applications*. Washington, DC: Island Press, 2011.

Daniels, Norman. *Am I My Parent's Keeper? An Essay on Justice between the Young and Old*. New York: Oxford University Press, 1990.

Darling, Frank Fraser, and John P. Milton. *Future Environments of North America*. Garden City, NY: Natural History Press, 1966.

Darroch, Jacqueline E., and Susheela Singh. *Estimating Unintended Pregnancies Averted by Couple-Years of Protection (CYP)*. New York: Guttmacher Institute, 2011.

Dasgupta, Aparajita, and Soumya Deb. "Telemedicine: A New Horizon in Public Health in India." *Indian Journal of Community Medicine* 33, no. 1 (2008): 3–8.

Davidson, Eric. *You Can't Eat GNP: Economics as if Ecology Mattered*. Cambridge, MA: Perseus Publishing, 2000.

de Beijer, Dorry. "Motherhood and New Forms of Reproductive Technology: Passive Source of Nutrition and Rational Consumer." In *Motherhood: Experience, Institution, Theology*, edited by Anne Carr and Elizabeth Schussler Fiorenza, 73–81. Edinburgh: T. and T. Clark, 1989.

de Crespigny, Lachlan, and Julian Savulescu. "Homebirth and the Future Child." *Journal of Medical Ethics* 40, no. 12 (2014): 807–12.

Denny, Colleen C., and Ezekiel J. Emanuel. "US Health Aid beyond PEPFAR: The Mother and Child Campaign." *Journal of the American Medical Association* 300, no. 17 (2008): 2048–51.

Department of Health and Human Services. *Public Health Notification: PVC Devices Containing the Plasticizer DEHP*. Rockville, MD: Food and Drug Administration, July 12, 2002.

Department of Health and Human Services, Departmental Appeals Board Appellate Division NCD 140.3. Transsexual Surgery Docket no. A-13–87, Decision no. 2576. May 30, 2014.

Derrett, Sarah, Charlotte Paul, and Jenny M. Morris. "Waiting for Elective Surgery: Effects on Health-Related Quality of Life." *International Journal for Quality in Health Care* 11, no. 1 (1999): 47–57.

De Vitoria, Francisco. "On the American Indians." In *Vitoria Political Writings: Cambridge Texts in the History of Political Thought*, edited by Anthony Pagden and Jeremy Lawrance, 231–92. Cambridge: Cambridge University Press, 1991.

De Vries, Annelou L. C., Jenifer K. McGuire, Thomas D. Steensma, Eva C. F. Wagenaar, Theo A. H. Doreleijers, and Peggy T. Cohen-Kettenis. "Young Adult Psychological Outcome after Puberty Suppression and Gender Reassignment." *Pediatrics* 134, no. 4 (2014): 696–704.

Di Paola, Marcello, and Mirko Daniel Garasic. "The Dark Side of Sustainability: On Avoiding, Engineering, and Shortening Human Lives in the Anthropocene." *Rivista di Studi sulla Sostenibilità* 3, no. 2 (2013): 59–81.

Di Tella, Rafael, and Robert MacCulloch. "Gross National Happiness as an Answer to the Easterlin Paradox?" *Journal of Development Economics* 86, no. 1 (2008): 22–42.

Djordjevic, Miroslav L., Marta R. Bizic, Dragana Duisin, Mark-Bram Bouman, and Marlon Buncamper. "Reversal Surgery in Regretful Male-to-Female Transsexuals after Sex Reassignment Surgery." *Journal of Sexual Medicine* 13, no. 6 (2016): 1000–1007.

Dolan, Paul, Rebecca Shaw, Aki Tsuchiya, and Alan Williams. "QALY Maximisation and People's Preferences: A Methodological Review of the Literature." *Health Economics* 14, no. 2 (2005): 197–208.

Donchin, Anne. "In Whose Interest? Policy and Politics in Assisted Reproduction." *Bioethics* 25, no. 2 (2011): 92–101.

Donnelly, Laura. "NHS Provokes Fury with Indefinite Surgery Ban for Smokers and Obese." *Telegraph*, October 17, 2017.

Donnelly, Laura. "Single Women Should Not Get Free IVF, Say Ethics Experts." *Telegraph*, July 25, 2014.

Donohoe, Martin, and Gordon Schiff. "A Call to Service: Social Justice Is a Public Health Issue." *Virtual Mentor: American Medical Association Journal of Ethics* 16, no. 9 (2014): 699–707.

Dreaper, Jane. "The One Dollar Contraceptive Set to Make Family Planning Easier." *BBC News*, November 15, 2014.

Dugger, Celia W. "Clinton Gets Five Companies to Reduce the Cost of AIDS Tests." *New York Times*, January 15, 2004.

Dworkin, Gerald. "Paternalism." In *Intervention and Reflection: Basic Issues in Medical Ethics*, 8th ed., edited by Ronald Munson, 125–34. Australia: Thompson, 2008.

Dwyer, James. "How to Connect Bioethics and Environmental Ethics: Health, Sustainability, and Justice." *Bioethics* 23, no. 9 (2009): 497–502.

Easterlin, Richard A. "Will Raising the Incomes of All Increase the Happiness of All?" *Journal of Economic Behavior & Organization* 27, no. 1 (1995): 35–47.

Ehrlich, Paul. "Ecoethics: Now Central to All Ethics." *Bioethical Inquiry* 6 (2009): 417–36.

Ehrlich, Paul R., and John P. Holdren. "Impact of Population Growth." *Science* 171, no. 3977 (1971): 1212–17.

Elgin, Duane. *Voluntary Simplicity: Toward a Way of Life That Is Outwardly Simple, Inwardly Rich.* New York: Quill, 1993.

Egger, Garry. "Personal Carbon Trading: A Potential 'Stealth Intervention' for Obesity Reduction?" *Medical Journal of Australia* 187, no. 3 (2007): 185–87.

European Commission, Emissions Database for Global Atmospheric Research. "CO2 Time Series 1990–2015 per Region/Country." http://edgar.jrc.ec.europa.eu.

Eyal, Nir, and Till Bärnighausen. "Precommitting to Serve the Underserved." *American Journal of Bioethics* 12, no. 5 (2012): 23–34.

Fassin, Didier, and Richard Rechtman. *The Empire of Trauma: An Inquiry into the Condition of Victimhood.* Translated by Rachel Gomme. Princeton, NJ: Princeton University Press, 2009.

Fildes, Alison, Judith Charlton, Caroline Rudisill, Peter Littlejohns, A. Toby Prevost, and Martin C. Gulliford. "Probability of an Obese Person Attaining Normal Body Weight: Cohort Study Using Electronic Health Records." *American Journal of Public Health* 105, no. 9 (2015): e54–59.

Fishel, Jason Lee. "The Green Staff of Asclepius: Envisioning Sustainable Medicine." PhD diss., University of Tennessee, Knoxville, 2014.

Fleck, Leonard M. *Just Caring: Health Care Rationing and Democratic Deliberation.* New York: Oxford University Press, 2009.

Freyer, Felice J. "Specialty Drugs Transform Lives—but at a Cost." *Boston Globe*, July 21, 2014.

Fuller, Jon, and James Keenan. "Educating in a Time of HIV/AIDS: Learning from the Legacies of Human Rights, the Common Good, and the Works of Mercy." In *Opening Up: Speaking Out in the Church*, edited by Julian Filochowski and Peter Stanford, 95–113. London: Darton, Longman & Todd, 2005.

Gallagher, James. "First Human Eggs Grown in Laboratory." *BBC News*, February 9, 2018.

Gallagher, James. "South Africans Perform First 'Successful' Penis Transplant." *BBC News*,

March 13, 2015.

Galvão, Jane. "Brazil and Access to HIV/AIDS Drugs: A Question of Human Rights and Public Health." *American Journal of Public Health* 95, no. 7 (2005): 1100–116.

Gibbons, Luz, José M. Belizán, Jeremy A. Lauer, Ana P. Betrán, Mario Merialdi, and Fernando Althabe. *The Global Numbers and Costs of Additionally Needed and Unnecessary Caesarean Sections Performed per Year: Overuse as a Barrier to Universal Coverage*: World Health Report. Geneva: World Health Organization, 2010.

Gill-Petersen, Julian. "The Value of the Future: The Child as Human Capital and the Neoliberal Labor of Race." *WSQ: Women's Studies Quarterly* 43, no. 1–2 (2015): 181–96.

Ginsburg, Paul B. *Wide Variation in Hospital and Physician Payment Rates Evidence of Provider Market Power*. Washington, DC: Center for Studying Health System Change, 2010.

Girod, Jennifer. "A Sustainable Medicine: Lessons from the Old Order Amish." *Journal of Medical Humanities* 23, no. 1 (2002): 31–42.

Gleick, Peter H., and Heather S. Cooley. "Energy Implications of Bottled Water." *Environmental Research Letters* 4, no. 1 (2009): 014009.

Goldemberg, José. "Leapfrog Energy Technologies." *Energy Policy* 26, no. 10 (1998): 729–41.

Gortmaker, Steven L., Boyd A. Swinburn, David Levy, Rob Carter, Patricia L. Mabry, Diane T. Finegood, Terry Huang, Tim Marsh, and Marjory L. Moodie. "Changing the Future of Obesity: Science, Policy, and Action." *Lancet* 378, no. 9793 (2011): 838–47.

Gostin, Lawrence O. "Public Health Law in an Age of Terrorism: Rethinking Individual Rights and Common Goods." *Health Affairs* 21, no. 6 (2002): 79–93.

Govindasamy, Bala, and Ken Caldeira. "Geoengineering Earth's Radiation Balance to Mitigate CO_2-Induced Climate Change." *Geophysical Research Letters* 27, no. 14 (2000): 2141–44.

Gracia, Clarisa R., Jorge J. E. Gracia, and Shasha Chen. "Ethical Dilemmas in Oncofertility: An Exploration of Three Clinical Scenarios." In *Oncofertility: Ethical, Legal, Social, and Medical Perspectives*, edited by Teresa Woodruff, Laurie Zoloth, Lisa Campo-Engelstein, and Sarah Rodriguez, 195–208. New York: Springer, 2010.

Grady, Denise. "Ebola Vaccine, Ready for Test, Sat on the Shelf." *New York Times*, October 23, 2014.

Groskop, Viv. "Do You Really Need a 'Mommy Makeover'?" *Guardian*, August 4, 2008.

Gwatkin, Davidson R., Abbas Bhuiya, and Cesar G. Victora. "Making Health Systems More Equitable." *Lancet* 364, no. 9441 (2004): 1273–80.

Halberstam, Judith. "F2M: The Making of Female Masculinity." In *Feminist Theory and the Body: A Reader*, edited by Janet Price and Margrit Shildrick, 125–33. New York: Routledge, 1999.

Hall, Amy Laura. *Conceiving Parenthood: American Protestantism and the Spirit of Reproduction*. Grand Rapids, MI: Eerdmans, 2008.

Hansen, James, Makiko Sato, Pushker Kharecha, David Beerling, Robert Berner, Valerie Masson-Delmotte, Mark Pagani, Maureen Raymo, Dana L. Royer, and James C. Zachos. "Target Atmospheric CO_2: Where Should Humanity Aim?" *Open Atmospheric Science Journal* 2 (2008): 217–31.

Hansen, Michèle, Lyn Colvin, Beverly Petterson, Jennifer J. Kurinczuk, Nicholas De Klerk, and Carol Bower. "Twins Born Following Assisted Reproductive Technology: Perinatal Outcome and Admission to Hospital." *Human Reproduction* 24, no. 9 (2009): 2321–31.

Hauskeller, Michael. "A Cure for Humanity: The Transhumanisation of Culture." *Trans-Humanities Journal* 8, no. 3 (2015): 131–47.

Hawken, Paul. *The Ecology of Commerce: A Declaration of Sustainability*. New York: Harper Business, 2010.

Hawley, Amos. "Ecology." In *International Encyclopedia of Population*, vol. 1, edited by John Ross, 159–63. New York: Free Press, 1982.

Heal, Geoffrey. "New Strategies for the Provision of Global Public Goods: Learning from International Environmental Challenges." In *Global Public Goods: International Cooperation in the 21st Century*, edited by Inge Kaul, Isabelle Grunberg, and Marc A. Stern, 220–39. New York: United Nations Development Programme, 1999.

Health Care Without Harm. *Healthier Hospitals Initiative*. Reston, VA: Health Care Without Harm, n.d.

Health Research for Action: UC Berkeley. "Telemedicine Social Franchising in Rural Uttar Pradesh, India." 2012. http://www.healthresearchforaction.org.

Hebden, Nicola. "Gay Couples 'Could Be Given Fertility Treatment.'" *The Local* (France), October 12, 2012.

Hensel, Bruce. "Doctor Claims Tongue Patch Can Help Shed Pounds." *NBC Los Angeles*, April 6, 2011.

Herbert, Geoff. "Buffalo Teachers Still Get Free Plastic Surgery Courtesy of Taxpayers." Syracuse.com, January 19, 2012.

Hewitt, Jacqueline K., Campbell Paul, Porpavai Kasiannan, Sonia R. Grover, Louise K. Newman, and Garry L. Warne. "Hormone Treatment of Gender Identity Disorder in a Cohort of Children and Adolescents." *Medical Journal of Australia* 196, no. 9 (2012): 578–81.

Heylens, Gunter, Els Elaut, Gerd Verschelden, and Greta De Cuypere. "Transgender Persons Applying for Euthanasia in Belgium: A Case Report and Implications for Assessment and Treatment." *Journal of Psychiatry* 18, no. 6 (2015): 1000347.

Hodgson, R. Eric, and Timothy C. Hardcastle. "South Africa: Where Have We Been?" In *ICU Resource Allocation in the New Millennium: Will We Say "No"?*, edited by David W. Crippen, 75–87. New York: Springer, 2013.

Hollenbach, David. "The Common Good as Participation in Community: A Theological/ Ethical Reflection on Some Empirical Issues." Lecture. Institute for Advanced Catholic Studies, University of Southern California, June 25–28, 2014, 15–18.

Holm, Soren. "Not Just Autonomy: The Principles of American Biomedical Ethics." *Journal of Medical Ethics* 21, no. 6 (1995): 332–38.

Horvitz, Eric, and Deirdre Mulligan. "Data, Privacy, and the Greater Good." *Science* (July 17, 2015): 253–55.

Howell, Joseph H., and William Frederick Sale. "Specifying the Goals of Medicine." In *Life Choices: A Hastings Center Introduction to Bioethics*, 2nd ed., edited by Joseph H. Howell and William Frederick Sale, 62–73. Washington, DC: Georgetown University Press, 2000.

"Intercytex Phase II Hair Multiplication Trial Update." *Tressless News*, September 25, 2007.

"Intercytex Seeks £15m on AIM to Combat Baldness." *Business Weekly U.K.*, January 27, 2006.

Jacobs, Lenworth M., and Robert J. Schwartz. "The Impact of Prospective Reimbursement on Trauma Centers: An Alternative Payment Plan." *Archives of Surgery* 121, no. 4 (1986): 479–83.

Jackson, Tim. "Live Better by Consuming Less: Is There a 'Double Dividend' in Sustainable Consumption?" *Journal of Industrial Ecology* 9, no. 1–2 (2005): 19–36.

Jackson, Tim. *Prosperity without Growth: Economics for a Finite Planet*. London: Earthscan, 2009.

Jacolin-Nackaerts, Myriam, and Jean Paul Clément. "La lutte contre l'obésité à l'école: Entre biopouvoir et individuation." *Lien social et Politiques*, no. 59 (2008): 47–60.

Jadoul, Pascale, Apolline Guilmain, J. Squifflet, Mathieu Luyckx, Raffaella Votino, Christine Wyns, and Marie-Madeleine Dolmans. "Efficacy of Ovarian Tissue

Cryopreservation for Fertility Preservation: Lessons Learned from 545 Cases." *Human Reproduction* 32, no. 5 (2017): 1046–54.

Jahr, Fritz. "'Bio+Ethik: Eine Umschau uber die ethischen Beziehungen des Menschen zu Tier und Pflanze." *Kosmos: Handweise fur Naturfreunde und Zentralblatt fur das naturwissenschaftliche Bildungs—und Sammelwesen*. Stuttgart, Germany: Kosmos, Gesellschaft der Naturfreunde, 1927.

Jahr, Fritz, and Hans-Martin Sass. "Bio-Ethics—Reviewing the Ethical Relations of Humans towards Animals and Plants." *JAHR-European Journal of Bioethics* 1, no. 2 (2010): 227–31.

Jalbert, Yves, and Lyne Mongeau. "Prévenir l'obésité: Un aperçu des programmes, plans d'action, stratégies et politiques sur l'alimentation et la nutrition." *Institut national de santé publique du Québec* (2006): 1–28.

Jameton, Andrew. "Outline of the Ethical Implications of Earth's Limits for Health Care." *Journal of Medical Humanities* 23, no. 1 (2002): 55.

Jenkins, Willis. *Ecologies of Grace: Environmental Ethics and Christian Theology*. Oxford: Oxford University Press, 2008.

Johanson, Richard, Mary Newburn, and Alison Macfarlane. "Has the Medicalisation of Childbirth Gone Too Far?" *BMJ* 324, no. 7342 (2002): 892–95.

Johnston, Josephine. "Why I Mostly Love ASB Bank's IVF Ad." Hastings Center Bioethics Forum, December 20, 2010. https://www.thehastingscenter.org/why-i-mostly-love-asb-banks-ivf-ad/.

Jones, Steven M., Heinz Feldmann, Ute Ströher, Joan B. Geisbert, Lisa Fernando, Allen Grolla, Hans-Dieter Klenk, Nancy Sullivan, Viktor Volchkov, Elizabeth Fritz, Kathleen Daddario, Lisa E. Hensley, Peter B. Jahrling, and Thomas W. Geisbert. "Live Attenuated Recombinant Vaccine Protects Nonhuman Primates against Ebola and Marburg Viruses." *Nature Medicine* 11, no. 7 (2005): 786–90.

Jonsen, Albert. *Responsibility in Modern Religious Ethics*. Washington, DC: Corpus Books, 1968.

Jonsen, Albert, Mark Siegler, and William Winslade. *Clinical Ethics: A Practical Approach to Ethical Decisions in Clinical Medicine*. 7th ed. New York: McGraw-Hill, 2010.

Jonsen, Albert R., Mark Siegler, and William J. Winslade. *Clinical Ethics: A Practical Approach to Ethical Decisions in Clinical Medicine*. 8th ed. New York: McGraw-Hill, 2015.

Kant, Immanuel. *On the Metaphysics of Morals and Ethics*. Translated by Thomas

Kingsmill Abbot. Overland Park, KS: Digireads Publishing, 2017.

Kass, Nancy. "Public Health Ethics: From Foundations and Frameworks to Justice and Global Public Health." *Journal of Law, Medicine and Ethics* 32, no. 2 (2004): 232–42.

Kaul, Inge, Isabelle Grunberg, and Mark Stern, eds. *Global Public Goods International Cooperation in the 21st Century*. Oxford: Oxford University Press, 1999.

Kennedy, Emily Huddart, Harvey Krahn, and Naomi Krogman. "Downshifting: An Exploration of Motivation, Quality of Life, and Environmental Practices." *Sociological Forum* 28, no. 4 (2013): 764–83.

Kerz, Thomas. "Germany: Where Are We Going?" In *ICU Resource Allocation in the New Millennium: Will We Say "No"?*, edited by David Crippen, 131–38. New York: Springer, 2013.

Kidder, Tracy. *Mountains beyond Mountains*. New York: Random House, 2009.

Kiisel, Ty. "You Are Judged by Your Appearance." *Forbes*, March 20, 2013.

Kime, Patricia. "DoD Spends $84M a Year on Viagra, Similar Meds." *Military Times*, February 13, 2015.

King, Louise P., and Janet Brown. "Clinical Case: Educating Patients as Medicine Goes Green." *Virtual Mentor: American Medical Association Journal of Ethics* 11, no. 6 (2009): 427–33.

Kmietowicz, Zosia. "Trust Defers Surgical Referrals for Patients to Lose Weight and Stop Smoking." *BMJ: British Medical Journal* (*Online*) 343 (2011).

Knowlton, Kim, Miriam Rotkin-Ellman, Linda Geballe, Wendy Max, and Gina M. Solomon. "Six Climate Change–Related Events in the United States Accounted for about $14 Billion in Lost Lives and Health Costs." *Health Affairs* 30, no. 11 (2011): 2167–76.

Koblinsky, Marjorie, Cheryl Moyer, Clara Calvert, James Campbell, Oona M. R. Campbell, Andrea B. Feigl, Wendy Graham, Laurel Hatt, Steve Hodgins, Zoe Matthews, Lori McDougall et al. "Quality Maternity Care for Every Woman, Everywhere: A Call to Action." *Lancet* 388, no. 10057 (2016): 2307–20.

Koch, Tom. *Thieves of Virtue: When Bioethics Stole Medicine*. Cambridge, MA: MIT Press, 2014.

Kotz, Deborah. "Hospitals Take Steps to Set Healthy Examples for Patients." *Boston Globe*, March 31, 2014.

Kugelman, Amir, Esther Inbar-Sanado, Eric S. Shinwell, Imad R. Makhoul, Meiron Leshem, Shmuel Zangen, Orly Wattenberg, Tanya Kaplan, Arieh Riskin, and

David Bader. "Iatrogenesis in Neonatal Intensive Care Units: Observational and Interventional, Prospective, Multicenter Study." *Pediatrics* 122, no. 3 (2008): 550–55.

Kukla, Rebecca. *Mass Hysteria: Medicine, Culture, and Mothers' Bodies*. New York: Rowman and Littlefield, 2005.

Launder, Brian, and J. Michael T. Thompson. *Geo-engineering Climate Change*. New York: Cambridge University Press, 2009.

Lavie, Carl, Paul McAuley, Timothy Church, Richard Milani, and Steven Blair. "Obesity and Cardiovascular Diseases: Implications Regarding Fitness, Fatness, and Severity in the Obesity Paradox." *Journal of the American College of Cardiology* 63, no. 14 (2014): 1345–54.

Lavie, Carl, Richard V. Milani, Hector O. Ventura, and Abel Romero-Corral. "Body Composition and Heart Failure Prevalence and Prognosis: Getting to the Fat of the Matter in the 'Obesity Paradox.'" *Mayo Clinic Proceedings* 85, no. 7 (2010): 605–8.

Lawrence, Anne. "Factors Associated with Satisfaction or Regret Following Male-to-Female Sex Reassignment Surgery." *Archives of Sexual Behavior* 32, no. 4 (2003): 299–315.

Lemoine, Marie-Eve, and Vardit Ravitsky. "Sleepwalking into Infertility: The Need for a Public Health Approach towards Advanced Maternal Age." *American Journal of Bioethics* 15, no. 11 (2015): 37–48.

Leopold, Aldo. *A Sand County Almanac and Sketches Here and There*. Oxford: Oxford University Press, 1968.

Letendre, Marie Catherine, and Joseph Tham. "Family and Healthcare Decision Making: Implications for Bioethics in China." *Studia Bioethica* 4, no. 3 (2011): 26–27.

Levy, Jenna. "In U.S., Uninsured Rate Sinks to 12.9." *Gallup News*, January 7, 2015.

Lyle, Katy, Louise Dent, Sally Bailey, Lynn Kerridge, Ian Roberts, and Ruairidh Milne. "Carbon Cost of Pragmatic Randomised Controlled Trials: Retrospective Analysis of Sample of Trials." *BMJ* 339 (2009): b4187.

Macpherson, Cheryl Cox. "Climate Change Is a Bioethics Problem." *Bioethics* 27, no. 6 (2013): 305–8.

Macpherson, Cheryl Cox, and Muge Akpinar-Elci. "Caribbean Heat Threatens Health, Well-being and the Future of Humanity." *Public Health Ethics* 8, no. 2 (2015): 196–208.

Malik, Khalid. *Human Development Report 2014: Sustaining Human Progress: Reducing Vulnerabilities and Building Resilience*. New York: United Nations Development Programme, 2014.

Martin, Joyce, Brady Hamilton, Michelle Osterman, Sally Curtin, and T. J. Mathews. "Births: Final Data for 2013." *National Vital Statistics Reports* 64, no. 1 (2015): 1–65.

Martini, Bruno. "The Anthropocene: Humankind as a Turning Point for Earth." *Astrobiology*, June 24, 2013.

Marx, Karl. *Capital: An Abridged Edition*. Edited by David McLellan. Oxford: Oxford World's Classics, 2008.

McKenny, Gerald P. "Enhancements and the Quest for Perfection." *Christian Bioethics* 5, no. 2 (1999): 99–103.

Mendola, Annette, Vicki Cannington, Lynnette Osterlund, and Caroline Vogel. "Worse Than Futile: Medically Non-Beneficial Treatment in the Setting of Complicated Grief." 11th Annual International Conference on Clinical Ethics and Consultation, New York City, May 21, 2015.

Miah, Andy, and Emma Rich. *The Medicalization of Cyberspace*. New York: Routledge, 2008.

Mill, John Stuart. *Utilitarianism* and *On Liberty*. Edited by Mary Warnock. London: Fontana Press, 1962.

Miller, Henry I. "Bespoke Drugs." *Forbes*, January 9, 2009.

Miller, Tracy. "Belgian Transsexual Dies by Euthanasia after Unsatisfactory Sex Change Operation." *New York Daily News*, October 2, 2013.

Mitchell, C. Ben, Edmund D. Pellegrino, Jean Bethke Elstain, John G. Kilner, and Scott B. Rae. *Biotechnology and the Human Good*. Washington, DC: Georgetown University Press, 2007.

Mor, Yechiel, and Joseph G. Schenker. "Ovarian Hyperstimulation Syndrome and Thrombotic Events." *American Journal of Reproductive Immunology* 72, no. 6 (2014): 541–48.

Morris, D. S., T. Wright, J. E. A. Somner, and A. Connor. "The Carbon Footprint of Cataract Surgery." *Eye* 27, no. 4 (2013): 495–501.

Morris, Lynn, and Nonhlanhla N. Mkhize. "Prospects for Passive Immunity to Prevent HIV Infection." *PLoS Medicine* 14, no. 11 (2017): e1002436.

Mujahid, Nisma, Yanke Liang, Ryo Murakami, Hwan Geun Choi, Allison S. Dobry, Jinhua Wang, Yusuke Suita, Qing Yu Weng, Jennifer Allouche, Lajos V. Kemeny, Andrea L. Hermann, Elisabeth M. Roider, Nathanael S. Gray, and David E. Fisher. "A UV-Independent Topical Small-Molecule Approach for Melanin Production in Human Skin." *Cell Reports* 19, no. 11 (2017): 2177–84.

Murphy, Julien S. "Is Pregnancy Necessary? Feminist Concerns about Ectogenesis." *Hypatia: Journal of Feminist Philosophy* 4, no. 3 (1989): 66–84.

Nash, James A. "Biotic Rights and Human Ecological Responsibilities." In *The Annual of the Society of Christian Ethics*, edited by Harlan Beckley, 137–62. Washington, DC: Georgetown University Press, 1993.

Nathan, Carl, and Otto Cars. "Antibiotic Resistance—Problems, Progress, and Prospects." *New England Journal of Medicine* 371, no. 19 (2014): 1761–63.

National Commission for the Protection of Human Subjects of Biomedical and Behavioral Research. *The Belmont Report: Ethical Principles and Guidelines for the Protection of Human Subjects of Research*. Washington, DC: US Government Printing Office, 1978.

National Health Service Sustainable Development Unit. *Saving Carbon, Improving Health: NHS Carbon Reduction Strategy for England*. London: NHS Sustainable Development Unit, 2009.

National Institute for Health and Care Excellence. "Offer Weight Loss Surgery to Obese People with Diabetes." *NICE*, November 27, 2014.

National Institute for Health Research. *The NIHR Carbon Reduction Guidelines* (England: NIHR, October 2010): 1–20.

National Quality Forum, National Quality Partners, and Antibiotic Stewardship Action Team. *National Quality Partners Playbook: Antibiotic Stewardship in Acute Care*. Washington, DC: National Quality Forum, 2016.

Nietzsche, Friedrich. *On the Genealogy of Morals*. Translated by Walter Kaufmann. New York: Vintage Books, 1967.

Northcott, Michael S. "The Concealments of Carbon Markets and the Publicity of Love in a Time of Climate Change." *International Journal of Public Theology* 4, no. 3 (2010): 294–313.

Nussbaum, Martha. "Aristotelian Social Democracy." In *Liberalism and the Good*, edited by R. Bruce Douglass, Gerald M. Mara, and Henry S. Richardson, 203–52. New York: Routledge, 1990.

Nussbaum, Martha C. "Human Functioning and Social Justice: In Defense of Aristotelian Essentialism." *Political Theory* 20, no. 2 (1992): 202–46.

O'Brien, Mary. *The Politics of Reproduction*. Boulder, CO: Westview Press, 1989.

O'Neill, Dan. "Gross Domestic Product." In *Degrowth: A Vocabulary for a New Era*. Edited by Giacomo D'Alisa, Federico Demaria, and Giorgos Kallis, 103–8. New York:

Routledge, 2014.

O'Rourke, Kevin. *A Primer for Health Care Ethics: Essays for a Pluralistic Society*. 2nd ed. Washington DC: Georgetown University Press, 2000.

Oakley, Ann. *The Captured Womb: A History of the Medical Care of Pregnant Women*. Oxford: Basil Blackwell, 1984.

Ofei, F. "Obesity—A Preventable Disease." *Ghana Medical Journal* 39, no. 3 (2005): 98.

Ogden, Cynthia L., Margaret D. Carroll, Brian K. Kit, and Katherine M. Flegal. "Prevalence of Childhood and Adult Obesity in the United States, 2011–2012." *Journal of the American Medical Academy* 311, no. 8 (2014): 806–14.

Ogden, Cynthia, Margaret D. Carroll, Cheryl D. Fryar, and Katherine M. Flegal. *Prevalence of Obesity among Adults and Youth: United States, 2011–2014*. Hyattsville, MD: US Department of Health and Human Services, Centers for Disease Control and Prevention, National Center for Health Statistics, 2015.

Olivennes, François. "Avoiding Multiple Pregnancies in ART Double Trouble: Yes a Twin Pregnancy Is an Adverse Outcome." *Human Reproduction* 15, no. 8 (2000): 1661–63.

Oreskes, Naomi, and Erik Conway. *Merchants of Doubt: How a Handful of Scientists Obscured the Truth on Issues from Tobacco Smoke to Global Warming*. London: Bloomsbury Publishing, 2010.

Organisation Mondiale de la Santé. "Stratégie mondiale pour l'alimentation, l'exercice physique et la santé." Fifty-seventh World Health Assembly, A57/9 (April 17, 2004): 1–24. http://apps.who.int.

Parens, Erik. "Is Better Always Good? The Enhancement Project." In *Enhancing Human Traits: Ethical and Social Implications*, edited by Erik Parens, 1–28. Washington, DC: Georgetown University Press, 1998.

Parens, Eric, and Adrienne Asch. "The Disability Rights Critique of Prenatal Genetic Testing: Reflections and Recommendations." *Special Supplement Hastings Center Report* 29 (1999): S1–S22.

Park, Kristin. "Stigma Management among the Voluntarily Childless." *Sociological Perspectives* 45, no. 1 (2002): 21–45.

Parker, Ryan. "Federal Judge Orders California Prison Inmate Be Granted Sexual Reassignment." *Los Angeles Times*, April 2, 2015.

Pascal, Jean, Hélène Abbey-Huguenin, and Pierre Lombrail. "Inégalités sociales de santé: Quels impacts sur l'accès aux soins de prévention?" *Lien social et Politiques–RIAC* 55, *La santé au risque du social* (Spring 2006): 115–24.

Patel, Sameer J., Adebayo Oshodi, Priya Prasad, Patricia Delamora, Elaine Larson, Theoklis Zaoutis, David A. Paul, and Lisa Saiman. "Antibiotic Use in Neonatal Intensive Care Units and Adherence with Centers for Disease Control and Prevention 12 Step Campaign to Prevent Antimicrobial Resistance." *Pediatric Infectious Disease Journal* 28, no. 12 (2009): 1047.

Peart, Karen N. "C-sections Linked to Breathing Problems in Preterm Infants." *Yale News*, February 10, 2010.

Pellegrino, Edmund. "Rationing Health Care: The Ethics of Medical Gatekeeping." *Journal of Contemporary Health Law and Policy* 2, no. 1 (1986): 23–46.

Pellegrino, Edmund D. "The Nazi Doctors and Nuremberg: Some Moral Lessons Revisited." *Annals of Internal Medicine* 127, no. 4 (1997): 307–8.

Pellegrino, Edmund D., and David C. Thomasma. *A Philosophical Basis of Medical Practice: Toward a Philosophy and Ethic of the Healing Professions*. New York: Oxford University Press, 1981.

Pellegrino, Edmund D., and David C. Thomasma. *The Virtues in Medical Practice.* New York: Oxford University Press, 1993.

Perry, Arthur W. *Straight Talk about Cosmetic Surgery*. New Haven, CT: Yale University Press, 2007.

Perry, Phil. "State of the Art and Science Greener Clinics, Better Care." *Virtual Mentor: American Medical Association Journal of Ethics* 16, no. 9 (2014): 726–31.

Peterson-Iyer, Karen. "Pharmacogenomics, Ethics, and Public Policy." *Kennedy Institute of Ethics Journal* 18, no. 1 (2008): 35–56.

Pianin, Eric, and Brianna Ehley. "Budget Busting U.S. Obesity Costs Climb Past $300 Billion a Year." *Fiscal Times*, June 19, 2014.

Picard, Frédéric, and Leonard Guarente. "Calorie Restriction—the *SIR2* Connection." *Cell* 120, no. 4 (2005): 473–82.

Pierce, Jessica, and Andrew Jameton. *The Ethics of Environmentally Responsible Health Care*. New York: Oxford University Press, 2004.

Pierce, Jessica, and Andrew Jameton. "Sustainable Health Care and Emerging Ethical Responsibilities." *Canadian Medical Association Journal* 164, no. 3 (2001): 365–69.

Potter, Van Rensselaer. *Bioethics: Bridge to the Future*. Englewood Cliffs, NJ: Prentice-Hall, 1971.

Potter, Van Rensselaer. "Bioethics: The Science of Survival." *Perspectives in Biology and Medicine* 14, no. 1 (1982): 127–53.

Potter, Van Rensselaer. *Global Bioethics: Building on the Leopold Legacy*. East Lansing: Michigan State University Press, 1988.

Powell-Dunford, Nicole, Amanda Cuda, Jeffrey Moore, Mark Crago, Amanda Kelly, and Patricia Deuster. "Menstrual Suppression for Combat Operations: Advantages of Oral Contraceptive Pills." *Women's Health Issues* 21, no. 1 (2011): 86–91.

Price, Frances. "The Management of Uncertainty in Obstetric Practice: Ultrasonography, In Vitro Fertilisation, and Embryo Transfer." In *The New Reproductive Technologies*, edited by Maureen McNeil, Ian Varcoe, and Steven Yearley, 123–53. London: Macmillan, 1990.

Purdy, Laura. "A Bioethics Perspective on Sex Reassignment Therapy." Unpublished manuscript (April 2015): 4–7. https://www.academia.edu/11995270.

Pym, Hugh. "Pfizer and Flynn Pharma Accused of Overcharging by CMA." *BBC News*, August 6, 2015.

Quinn, Gwendolyn P., Daniel K. Stearsman, Lisa Campo-Engelstein, and Devin Murphy. "Preserving the Right to Future Children: An Ethical Case Analysis." *American Journal of Bioethics* 12, no. 6 (2012): 38–43.

Rasmussen, Larry L. "Next Journey: Sustainability for Six Billion and More." In *Ethics for a Small Planet: New Horizons on Population, Consumption, and Ecology*, edited by Daniel Maguire and Larry L. Rasmussen, 67–140. New York: State University of New York Press, 1998.

Rawls, John. *Theory of Justice*. Cambridge, MA: Harvard University Press, 1971.

Reece, L. J., P. Sachdev, R. J. Copeland, M. Thomson, J. K. Wales, and N. P. Wright. "Use of Intragastric Balloons and a Lifestyle Support Programme to Promote Weight Loss in Severely Obese Adolescents: Pilot Study." *Appetite* 89 (2015): 305.

Reich, Warren T., ed. *The Encyclopedia of Bioethics*, vol. 1. New York: Macmillan, 1978.

Reich, Warren T. "The Word 'Bioethics': The Struggle Over Its Earliest Meanings." *Kennedy Institute of Ethics Journal* 5, no. 1 (1995): 19–34.

Reigstad, M., I. K. Larsen, T. Å. Myklebust, T. E. Robsahm, N. B. Oldereid, A. K. Omland, S. Vangen, L. A. Brinton, and R. Storeng. "Cancer Risk among Parous Women Following Assisted Reproductive Technology." *Human Reproduction* 30, no. 8 (2015): 1952–63.

Relman, Arnold S. "The New Medical-Industrial Complex." *New England Journal of Medicine* 303, no. 17 (1980): 963–70.

Resnik, David. *Environmental Health Ethics*. Cambridge: Cambridge University Press, 2012.

Reynolds, Gretchen. "The Stuttering Doctor's 'Monster Study.'" In *Ethics: A Case Study from Fluency*, edited by Robert Goldfarb, 1–12. San Diego: Plural Publishing, 2006.

Rich, Adrienne. *Of Woman Born: Motherhood as Institution and Experience*. New York: W.W. Norton, 1986.

Rich, Emma, John Evans, and Laura De Pian. "Children's Bodies, Surveillance and the Obesity Crisis." In *Debating Obesity: Critical Perspectives*, edited by Emma Rich, Lee Monaghan, and Lucy Aphramor, 139–63. New York: Palgrave Macmillan, 2010.

Richie, Cristina. "A Brief History of Environmental Bioethics." *Virtual Mentor: American Medical Association Journal of Ethics* 16, no. 9 (2014): 749–52.

Richie, Cristina. "A Queer, Feminist Bioethics Critique of Facial Feminization Surgery." *American Journal of Bioethics* 18, no. 12 (2018): 33–35.

Richie, Cristina. "Building a Framework for Green Bioethics: Integrating Ecology into the Medical Industry." *Health Care Ethics USA* 21, no. 4 (2013): 7–21.

Richie, Cristina. "Feminist Bioethics, Pornography, and the Reproductive Technologies Business." *Blog of IJFAB: the International Journal of Feminist Approaches to Bioethics*, October 5, 2015.

Richie, Cristina. "Reading Between the Lines: Infertility and Current Health Insurance Policies in the United States." *Clinical Ethics* 9, no. 4 (2014): 127–34.

Richie, Cristina. "Voluntary Sterilization for Childfree Women: Understanding Patient Profiles, Evaluating Accessibility, Examining Legislation." *Hastings Center Report* 43, no. 6 (2013): 36–44.

Richie, Cristina. "What Would an Environmentally Sustainable Reproductive Technology Industry Look Like?" *Journal of Medical Ethics* 41, no. 5 (2015): 383–87.

Risse, Guenter B. *Mending Bodies, Saving Souls: A History of Hospitals*. Oxford: Oxford University Press, 1999.

Roberts, Ian. "The NHS Carbon Reduction Strategy." *BMJ* 38, no. 7689 (2009): 248–49.

Roberts, Michelle. "Viagra Can Be Sold Over the Counter." *BBC News*, November 28, 2017.

Rosenthal, Elisabeth. "Is This a Hospital or a Hotel?" *New York Times*, September 21, 2013.

Rousseau, Jean-Jacques. *Discourse on the Origin of Inequality*. Translated by Donald A. Cress. Indianapolis, IN: Hackett, 1992.

Rowland, Christopher. "Hazards Tied to Medical Records Rush." *Boston Globe*, July 20, 2014.

Rushton, Christine. "Company Hikes Price 5,000% for Drug that Fights Complication of AIDS, Cancer." *USA Today*, September 18, 2015.

Sacitharan, Pradeep K., Sarah J. B. Snelling, and James R. Edwards. "Aging Mechanisms in Arthritic Disease." *Discovery Medicine* 14, no. 78 (2012): 345–52.

Santana, Danielly S., José Cecatti, Fernanda Surita, Carla Silveira, Maria Costa, João Souza, Syeda Mazhar, K. Jayaratne, Z. Qureshi, M. Sousa, and J. Vogel. "Twin Pregnancy and Severe Maternal Outcomes: The World Health Organization Multicountry Survey on Maternal and Newborn Health." *Obstetrics & Gynecology* 127, no. 4 (2016): 631–41.

Satcher, David. "The Legacy of the Syphilis Study at Tuskegee in African American Men on Health Care Reform Fifteen Years after President Clinton's Apology." *Ethics and Behavior* 22, no. 6 (2012): 486–88.

Schiffman, Mark, Philip E. Castle, Jose Jeronimo, Ana C. Rodriguez, and Sholom Wacholder. "Human Papillomavirus and Cervical Cancer." *Lancet* 370, no. 9590 (2007): 890–907.

Schumacher, Ernst. *Small Is Beautiful: Economics as if People Mattered.* New York: Harper and Row, 1973.

Schwenkenbecher, Anne. "Is There an Obligation to Reduce One's Individual Carbon Footprint?" *Critical Review of International Social and Political Philosophy* 17, no. 2 (2014): 168–88.

Scientific Committee on Emerging and Newly-Identified Health Risks. *Preliminary Opinion on the Safety of Medical Devices Containing DEHP-Plasticized PVC or Other Plasticizers on Neonates and Other Groups Possibly at Risk.* Luxembourg: European Union, 2014.

Sen, Amartya. "Equality of What?" Tanner Lectures on Human Values. Stanford University, May 22, 1979.

Sénat de Belgique. "La loi du 6 juillet 2007 relative à la procréation médicalement assistée et à la destination des embryons surnuméraires et gametes." *Moniteur belge* 38575 (2007).

Seppelt, Ian M. "Australia: Where Are We Going?" In *ICU Resource Allocation in the New Millennium: Will We Say "No"?*, edited by David Crippen, 107–12. New York: Springer, 2013.

Seppelt, Ian M. "Australia: Where Have We Been?" In *ICU Resource Allocation in the New Millennium: Will We Say "No"?*, edited by David Crippen, 3–10. New York: Springer, 2013.

Shontell, Alyson. "If You Look Like This, Your Pay Check Will Be Higher Than Average." *Business Insider*, February 26, 2011.

Shue, Henry. "Subsistence Emissions and Luxury Emissions." *Law & Policy* 15, no. 1 (1993): 39–60.

Stoop, Dominic, Ana Cobo, and Sherman Silber. "Fertility Preservation for Age-Related Fertility Decline." *Lancet* 384, no. 9950 (2014): 1311–19.

Streat, Stephen. "New Zealand: Where Have We Been?" In *ICU Resource Allocation in the New Millennium: Will We Say "No"?*, edited by David W. Crippen, 65–73. New York: Springer, 2013.

Sulmasy, Daniel P. "Dignity and Bioethics: History, Theory, and Selected Applications." In *Human Dignity and Bioethics: Essays Commissioned by the President's Council on Bioethics*, edited by Adam Schulman, 469–501. Washington, DC: President's Council on Bioethics, 2008.

Sulmasy, Daniel, and Beverly Moy. "Debating the Oncologist's Role in Defining the Value of Cancer Care: Our Duty Is to Our Patients." *Journal of Clinical Oncology* 32, no. 36 (2014): 4039–41.

Suresh, Gautham, Jeffrey Horbar, Paul Plsek, James Gray, William Edwards, Patricia Shiono, Robert Ursprung, Julianne Nickerson, Jerold Lucey, and Donald Goldmann. "Voluntary Anonymous Reporting of Medical Errors for Neonatal Intensive Care." *Pediatrics* 113, no. 6 (2004): 1609–18.

Szasz, Thomas. *The Medicalization of Everyday Life: Selected Essays*. Syracuse, NY: Syracuse University Press, 2007.

Taylor, Carol, and Robert Barnet. "Hand Feeding: Moral Obligation or Elective Intervention?" *Health Care Ethics USA* 22, no. 2 (2014): 12–23.

Thorniley, Andrew. "United Kingdom: Where Are We Going?" In *ICU Resource Allocation in the New Millennium: Will We Say "No"?*, edited by David Crippen, 177–83. New York: Springer, 2013.

Tiefer, Leonore. "The Viagra Phenomena." *Sexualities* 9, no. 3 (2006): 273–94.

Transgender Law Center. "10 Tips for Working with Transgender Patients." 2016. http:// transgenderlawcenter.org.

Tomson, Charles, Robert Foley, Q. Li, David Gilbertson, Jay Xue, and Allan Collins. "Race and End-Stage Renal Disease in the United States Medicare Population: The Disparity Persists." *Nephrology Carlton* 13, no. 7 (2008): 651–56.

Townes, Emilie M. *Breaking the Fine Rain of Death: African American Health Issues and a Womanist Ethic of Care*. New York: Continuum, 1998.

Trost, Landon, and Robert Brannigan. "Oncofertility and the Male Cancer Patient."

Current Treatment Options in Oncology 13, no. 2 (2012): 146–60.

Tul, Natasa, Miha Lucovnik, Ivan Verdenik, Mirjam Druskovic, Ziva Novak, and Isaac Blickstein. "The Contribution of Twins Conceived by Assisted Reproduction Technology to the Very Preterm Birth Rate: A Population-Based Study." *European Journal of Obstetrics & Gynecology and Reproductive Biology* 171, no. 2 (2013): 311–13.

United Church of Christ Commission on Racial Justice. *Toxic Wastes and Race in the United States: A National Report on the Racial and Socio-Economic Characteristics of Communities with Hazardous Waste Sites.* New York: United Church of Christ, 1987.

United Kingdom Legislation. "Climate Change Act 2008." https://www.legislation.gov.uk.

United Nations. *The Millennium Development Goals Report.* New York: United Nations, 2013.

United Nations. *Universal Declaration of Human Rights.* Geneva: United Nations, 1948.

United Nations Development Programme. *Human Development Report 2016: Human Development for Everyone.* New York: United Nations Development Programme, 2016.

United Nations Division of Economic and Social Affairs. *World Population Prospects: The 2012 Revision.* Vol. 1, *Comprehensive Tables.* New York: United Nations, 2013.

United Nations Educational, Scientific and Cultural Organization. *Universal Declaration of Animal Rights.* Paris: UNESCO, 1978.

United Nations Programme on HIV and AIDS. *Gap Report.* Geneva: UNAIDS, 2014.

United Nations Statistics Division Millennium Development Goals Indicators. "Carbon Dioxide Emissions (CO_2), Metric Tons of CO_2 per Capita (CDIAC)." https://millenniumindicators.un.org/unsd/mdg/SeriesDetail.aspx?srid=751.

United States Census Bureau. "Census Bureau Projects U.S. and World Populations on New Year's Day." United States Census Bureau, December 29, 2014.

United States Department of Commerce, International Trade Administration. *2016 Top Markets Report: Pharmaceuticals* (2016). https://www.trade.gov/topmarkets/.

United States Department of Commerce, National Oceanic and Atmospheric Administration Earth System Research Laboratory, Global Monitoring Division. "Trends in Atmospheric Carbon Dioxide: Recent Monthly Average Mauna Loa CO_2." https://www.esrl.noaa.gov.

United States Food and Drug Administration. *Approved Drug Products with Therapeutic Equivalence Evaluations.* 35th ed. Silver Spring, MD: U.S. Food and Drug Administration, 2015.

United States Food and Drug Administration. "Birth Control." https://www.fda.gov.

United States Food and Drug Administration. "FDA Approves First-of-Kind Device to Treat Obesity." *ScienceDaily*, January 29, 2015.

Veatch, Robert M. "How Many Principles for Bioethics?" In *Principles of Health Care Ethics*, 2nd ed., edited by Richard Edmund Ashcroft, Angus Dawson, Heather Draper, and John McMillan, 43–50. West Sussex, England: John Wiley & Sons, 2007.

Venteicher, Wes. "Insurer Lowers Cost of HIV Treatments after Discrimination Complaints." *Chicago Tribune*, March 27, 2015.

Vicini, Andrea. "Is Transhumanism a Helpful Answer to Contemporary Bioethical Challenges?" Lecture. Ethics Grand Rounds, University of Texas Southwestern Medical Center, Dallas, Texas, March 11, 2014. https://utswmed-ir.tdl.org/utswmed-ir/.

Viers, Hillary Wicai. "What Is Democratic Deliberation? A Q&A with Bioethics Commission Chair Amy Gutmann." *Blog of the Presidential Commission for the Study of Bioethical Issues*, September 10, 2014. https://bioethicsarchive.georgetown.edu.

Voigt, Kristin, and Harald Schmidt. "Gastric Banding: Ethical Dilemmas in Reviewing Body Mass Index Thresholds." *Mayo Clinic Proceedings* 86, no. 10 (2011): 999–1001.

Wakefield, Jane. "First 3D-Printed Pill Approved by US Authorities." *BBC News*, August 4, 2015.

Walker, Francis O. "Essay Cultivating Simple Virtues in Medicine." *Neurology* 65, no. 10 (2005): 1678–80.

Wallace, Sumer Allensworth, Kiara L. Blough, and Laxmi A. Kondapalli. "Fertility Preservation in the Transgender Patient: Expanding Oncofertility Care beyond Cancer." *Gynecological Endocrinology* 30, no. 12 (2014): 868–71.

Wang, Y. Claire, Klim McPherson, Tim Marsh, Steven L. Gortmaker, Martin Brown. "Health and Economic Burden of the Projected Obesity Trends in the USA and the UK." *Lancet* 378, no. 9793 (2011): 815–25.

Warren, Charlotte E., Pooja Sripad, Annie Mwangi, Charity Ndwiga, Wilson Liambila, and Ben Bellows. "'Sickness of Shame': Investigating Challenges and Resilience among Women Living with Obstetric Fistula in Kenya." In *Global Perspectives on Women's Sexual and Reproductive Health across the Lifecourse*, edited by Shonali Choudhury, Jennifer Toller Erausquin, and Mellissa Withers, 91–109. New York: Springer, 2018.

Wax, Randy S. "Canada: Where Are We Going?" In *ICU Resource Allocation in the New Millennium: Will We Say "No"?*, edited by David Crippen, 123–29. New York: Springer, 2013.

Whitty, Julia. "Diagnosing Health Care's Carbon Footprint." *Mother Jones*, November 10, 2009.

Wibulpolprasert, Suwit, and Paichit Pengpaibon. "Integrated Strategies to Tackle the Inequitable Distribution of Doctors in Thailand: Four Decades of Experience." *Human Resources for Health* 1, no. 1 (2003): 12.

Wilson, Jonathan, Ian Woods, Jayne Fawcett, Rebecca Whall, Wendy Dibb, Chris Morris, and Elizabeth McManus. "Reducing the Risk of Major Elective Surgery: Randomised Controlled Trial of Preoperative Optimization of Oxygen Delivery." *BMJ* 318, no. 7191 (1999): 1099–1103.

World Health Organization. *Density of Doctors, Nurses and Midwives in the 49 Priority Countries*. Geneva: WHO Global Atlas of the Health Workforce, 2010.

World Health Organization. "Density of Physicians (total number per 1000 population)." http://www.who.int.

World Health Organization. *Global Health Risks: Mortality and Burden of Diseases Attributable to Selected Major Risks*. Geneva: WHO Press, 2009.

World Health Organization. "Health Service Coverage: Data by Country: Births Attended by Skilled Health Personnel." http://www.who.int.

World Health Organization. *Health Workforce*. Geneva: WHO Global Atlas of the Health Workforce, 2010.

World Health Organization. "Maternal Mortality." http://www.who.int.

World Health Organization. *Preamble to the Constitution of the World Health Organization as Adopted by the International Health Conference, June 19–22, 1946*. Official Records of the World Health Organization, no. 2. New York: World Health Organization, 1948.

Wynia, Matthew. "Advocate as a Doctor or Advocate as a Citizen?" *Virtual Mentor: American Medical Association Journal of Ethics* 16, no. 9 (2014): 694–98.

Yeo, Michael. "Toward an Ethic of Empowerment for Health Promotion." *Health Promotion International* 8, no. 3 (1993): 225–35.

Young, Lindsay, Phil Schumm, Leigh Alon, Alida Bouris, Matthew Ferreira, Brandon Hill, Aditya S. Khanna, Thomas W. Valente, and John A. Schneider. "PrEP Chicago: A Randomized Controlled Peer Change Agent Intervention to Promote the Adoption of Pre-Exposure Prophylaxis for HIV Prevention among Young Black Men Who Have Sex With Men." *Clinical Trials* 15, no. 1 (2018): 44–52.

Zhong, Wenjun, Hilal Maradit-Kremers, Jennifer L. St. Sauver, Barbara P. Yawn, Jon O. Ebbert, Véronique L. Roger, Debra J. Jacobson, Michaela E. McGree, Scott M. Brue, and Walter A. Rocca. "Age and Sex Patterns of Drug Prescribing in a Defined American Population." *Mayo Clinic Proceedings* 88, no. 7 (2013): 697–707.

Zimbardo, Philip G. "On the Ethics of Intervention in Human Psychological Research: With Special Reference to the Stanford Prison Experiment." *Cognition* 2, no. 2 (1973): 243–56.

Zoloth, Laurie, and Alyssa Henning. "Bioethics and Oncofertility: Arguments and Insights from Religious Traditions." In *Oncofertility: Ethical, Legal, Social, and Medical Perspectives*, edited by Teresa Woodruff, Laurie Zoloth, Lisa Campo-Engelstein, and Sarah Rodriguez, 261–78. New York: Springer, 2010.

Index